MATHEMATICAL PHYSICS

A Popular Introduction

FRANCIS BITTER

Illustrations by
Kenneth Crook

DOVER PUBLICATIONS, INC.
Mineola, New York

Bibliographical Note

This Dover edition, first published in 2004, is an unabridged republication of the work originally titled *Mathematical Aspects of Physics: An Introduction,* originally published in 1963 by Anchor Books, Garden City, New York, as part of the Science Study Series.

International Standard Book Number: 0-486-43501-6

Manufactured in the United States of America
Dover Publications, Inc., 31 East 2nd Street, Mineola, N.Y. 11501

Contents

1. PATTERNS BEYOND LIFE 1

The Excitement of Physics. The Spectroscope. The Hydrogen Spectrum. The Significance of Physical Measurement.

2. DATA 25

Fact Gathering. Raw Data. The Path of Scientific Investigation. The Motion of the Earth around the Sun. The Motion of Venus. The Motion of Mercury.

3. ANALYSIS 67

Mathematics–The Language of Physics. Mechanical Vibrations. The Description of Oscillations. Simple Harmonic Motion. The Descriptive Elements of Simple Harmonic Shapes. A New Way of Describing Curves. The Geometry of the Laws of Motion. The Geometry of Oscillatory Motion. Determination of a Period of an Oscillation.

4. EXPERIMENTATION 129

Magnetic Fields. The Fields of Coils. Variables in Designing a Coil. Make-believe and Reality.

Appendix 163

Index 181

MATHEMATICAL PHYSICS

A Popular Introduction

CHAPTER 1

Patterns Beyond Life

One can be bored by almost anything, but it is quite a trick. It requires real indolence, the shutting off of all the senses, a state of more or less complete mental inactivity. Conversely, one can be excited and stimulated by almost anything, but that is even more of a trick. That happy state requires a true and constant awareness, using all one's senses. It requires mental agility, thinking ahead, anticipation. It is possible only in the midst of some challenging activity, where the outcome depends on one's actions, as in building a bridge, or making love, or driving a fast car.

Physics is boring only if one doesn't understand what it is all about–like listening to a speech in a language that one does not know. But in fact physics is a most exciting activity. It has to do with understanding, and it is most remarkable in that such great depth of understanding seems to be possible. When driving a car, it is important to know the answer to such questions as: Where is the ignition switch? Where is the accelerator? Where are the brakes? How do you turn on the lights? It is important to be familiar with its vibrations, the swing and the responses of the car, to know what the significance of unusual sounds and smells and sights may be. There are comparable items to be taken into consideration by a physicist when he is in the driver's seat in his professional work.

The Excitement of Physics

One thing you can say about physics is that it gives human beings important knowledge about the world in which they live. It is the basis of almost everything we make and use–from airplanes to compass needles, from television to toy balloons. But all this is only incidental. The real excitement is that physics tells us about our origins, our surroundings, and our destiny. What existed in the past, before we were born, long long before we were born? What is this world around us now? How is it made? And finally, how long will it last? How will it change?

Throughout history man has wanted answers to his questions. At first these answers were in the form of dramatic myths–fictions that we no longer believe because we have learned to "see" better, into the past, into the future, into the sky, into the atom. And what is it that man has learned to see with this new vision of his? Patterns, designs, symmetries, motions, changes interwoven with each other but more intricate and varied, more meaningful than the orderliness of man's inventions. Let us follow a train of thought to illustrate the magnitude of these differences.

When we look at a design or pattern, or when we examine some piece of reality, particularly if it is unfamiliar, we tend to ask ourselves:

How was it made?
Why?
What is it?
How does it work?

When we look at the scene in Fig. 1, even though

Fig. 1.

we may never have been there, we do not have to ask these questions—we feel that we know many of the answers. Within the framework available to him, the framework of earth and air, sun and rain, stone and wood, man has made this scene—to live in, to work in, to worship in. We feel that we know something of the character of the people who built this, and of the physical conditions of their environment there. It is a town, and its life functions according to a framework of customs and laws, factories and stores, schools and books, banks and farms that are familiar to us. It is man-made.

Now if we come nearer and examine a part of this scene, for example the church, we may see something like the pattern of Fig. 2. Again, at a glance, we feel

Fig. 2.

we know. It is man-made. It is a church for worship, for marriages, baptisms, funerals. It is for contemplation and prayer.

But perhaps we should begin to feel some hesitancy in suggesting that our church is altogether man-made. Surely men and women decorated and furnished it, put together the stone and glass, and drew the plans according to which the whole was assembled, but there is also clearly an inspiration, a motive, a drive which acted through the men and women who made it. Surely we are all aware that new ideas "come to us"—we don't know how to "make" them. Somehow the universe is so constituted that the human beings on the earth are capable of being "inspired" to build, to make, to look, and to perceive.

Perhaps some of you still feel that this is quite ordinary. There is nothing particularly inspired about spreading a canvas over a pole to get out of the rain. Nor in sawing up wood and tacking boards together to make a more solid house. Nor is any great inspiration needed to use stone, carefully cut and piled up, to make an even more solid structure. So let us look a little closer.

How about that figure kneeling in the church? What is that really? No more than clothes, draped over skin and flesh, held up by a skeleton? An intricately jointed pattern of bones like that in Fig. 3? How was that made? And what for? And how does

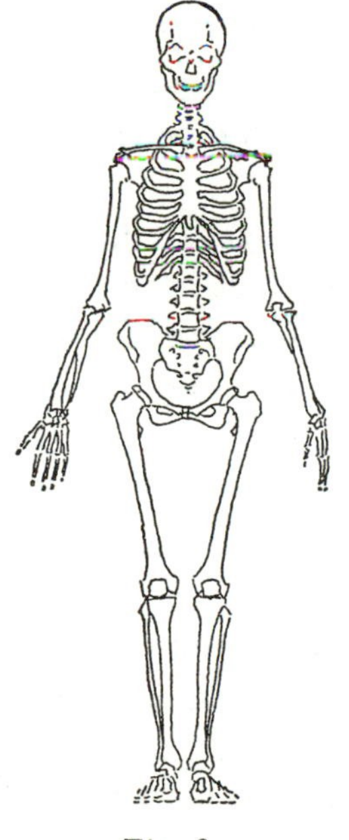

Fig. 3.

it work? Well, in a sense we can say that its father and mother made this skeleton, but when it was born it looked very different. Probably none of the atoms actually in that skeleton were ever in its mother's womb. They were assembled and cooked and fed into canals in the skin and flesh, and then finally drifted into place. In a word the skeleton grew. So, the most we can say is that the father and mother assembled the beginnings, and started it on its way with the first breath of life and the first heartbeats. From then on it was caught up in a stream of life that moved it along. We ought to know all about it. To all intents and purposes that skeleton is you or me. How was it made? Why? What is it? How does it function? We have one kind of information about this. We have experienced it. The broken leg grows together again. The cut finger heals. The skeleton grows, walks around, dances, sleeps. Why? Again, how does it function? Some of us add to our inner subjective feeling by looking, studying, experimenting, analyzing, searching–doctors, biologists, psychologists, lawyers, teachers, priests . . .

But we are still only part way along that train of thought. How about that beam of light shining down near that kneeling figure in the church? What is it that makes the tiny specks of dust (Fig. 4) in the air appear to shine as brightly as a milky way of stars? If we stop there, and merely notice that the beam lights up the dust spots and carpet in the aisle, and one end of two of the pews, and casts shadows, it soon loses interest for us. There is nothing more to "see." But are we really out of the interesting part of the world when we leave the town, and the church, and the human being, and enter into the sunlight, or is this

Fig. 4.

just another part of the same world? A part of the same life that builds and drives the skeleton (you and me)? The same life that inspires us to build the church or make life in a town? The answer depends on what we see when we look. Is it all evidently a part of the same scheme of things? Is the pattern equally deep and mysterious and complex? Or is it just "dead matter," to be used and kicked around—bricks and dust?

Two things are important to realize when we try to look deeply into the physical world around us. The first is that we do not see very far with our unaided senses. These were designed primarily so that we would go on living. We are learning how to use them in conjunction with our brains. To "see," we need microscopes and telescopes to help our eyes. We need magnetic and electric fields to do things that we cannot do with our hands and fingers. We need electronic devices like oscilloscopes and computers to follow the flashes and zip through the calculations for which our senses and brains are too sluggish. To "look at" the beam of light more deeply we need a prism, or a grating—a device for spreading out the inner structure, the pattern of wave lengths only millionths of an inch long that make up the light. With our eyes we see only the reds and greens and browns, the patterns of color and shade. But there is more to light than that.

The second thing that we shall find is that we can describe most of what we "see" only by numbers. Qualities have to do with our senses and with our feelings. The results of measurements with tools are best described in numbers. How big? How hot? How strong? Even color is described in numbers.

The Spectroscope

So, bearing these two things in mind, the need for tools and the need for numbers, let us look into the sunlight. Any large, polished piece of glass with corners can be used to break it up into its various colors. The pattern—we call it the spectrum—may be aesthetically pleasing, like a rainbow, but it does not tell us very much. To make real progress, we need a specially designed instrument: a spectroscope which breaks up the light into its various components in a measurable way.

Many readers know that light is an electromagnetic wave, and that different wave lengths seem to us to have different colors. The wave lengths of visible light range roughly from 0.00008, or 8×10^{-5}, centimeter in the red to 0.00004, or 4×10^{-5}, cm for the violet. (When we measure things in the material world, we must expect to use very different units from those made for our daily use, and especially we must appreciate exponents. Thus one centimeter is about the distance between these lines | |. One-tenth of a cm, or 0.1 cm, or 10^{-1} cm is about the distance between these lines ||. One hundredth of a cm, or 0.01 cm, or 10^{-2} cm is about the width of this line |, and so on.) By methods that need not concern us, these very short lengths can be measured, and with great accuracy. The details of the measuring process might be called technicalities. We are interested here only in the results. What is the sunlight like when we look deeply into it with an instrument designed to display spectra, a spectroscope?

Very few readers, probably, have thought of estimating how good a spectroscope would be needed to show the details of sunlight, or even what we mean by a good spectroscope or a bad one. Going back to our thoughts about a car, we might notice that different cars are good for different purposes, Cadillacs, Volkswagens, Isettas, Jeeps, trucks, earth-movers, etc., and that we can specify what we want without knowing very much of their detailed construction. Choosing a car is an exciting business, because it will be possible for us to do things with it that we could not do without it. So with spectroscopes.

For our purposes we might think of a spectroscope as an instrument which bends light of different wave lengths by different amounts. Our problem is to consider how accurately this is done, whether in the process of bending there may not also be a certain amount of mixing of adjacent wave lengths. The situation is very similar to tuning a radio. On the dial is a scale of wave lengths, or frequencies, and by turning the dial we can tune in on various stations broadcasting on different frequencies. While the readings on the dial may be perfectly correct, we know that confusion arises when the broadcast frequencies are too close together–the radio cannot separate them. There is a certain limit to the frequencies that the instrument can resolve, and, as you turn the dial, this is essentially the frequency range over which you can hear the signal from a single station.

The situation in a spectroscope is very similar. We show in Fig. 5a the light from a source focused by two lenses. In Fig. 5b we have added a prism which bends different waves by different amounts; it forms a red image of the candle using the red components of its light at a different place from the violet image. Between these two is a continuous range of images on different wave lengths. We may expect these images to overlap by an amount depending on their size. So, in order to reduce overlap, we should keep the object, and the image, small, at least in the plane of the paper in the sketch. A spectroscope used to examine sunlight, for instance, would have a narrow slit whose images, called lines, or spectral lines in different wave lengths would overlap as little as possible. This is shown in Fig. 5c. Lenses focus an image of the sun on a slit and of the slit on a screen. These images will necessarily have a certain size, and the resolving power of the spectroscope is determined by the range of wave lengths covered by the image of the slit. The smallest size to which the designer of spectroscopes actually can hold an image of the slit is now a crucial point. He cannot simply make the slit narrower and narrower. For large slits narrowing the slit makes the image smaller, but finally there comes a point at which the fuzzy edges of the image

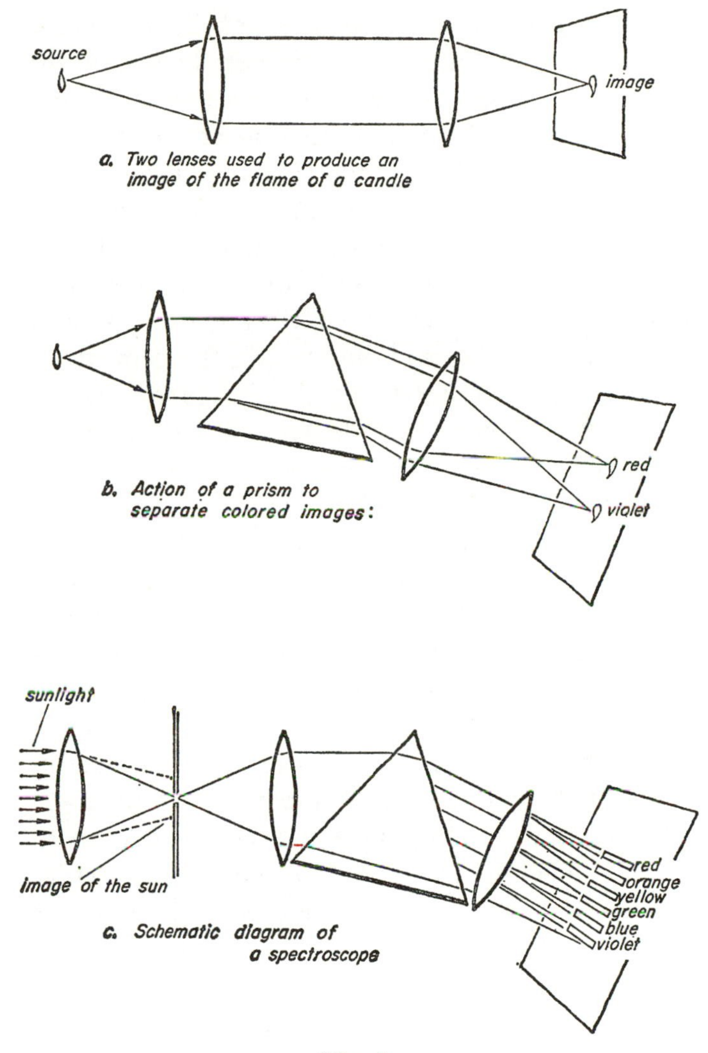
source
image
a. Two lenses used to produce an image of the flame of a candle
red
violet
b. Action of a prism to separate colored images:
sunlight
image of the sun
red
orange
yellow
green
blue
violet
c. Schematic diagram of a spectroscope

Fig. 5.

of the slit remain, and merely get fainter as the slit is further narrowed.

But this worry is in the province of the instrument designer. We shall merely specify here how good the spectroscope is insofar as resolving power is concerned. On the right of Fig. 5c is a screen on which there are all the colors of the rainbow. These correspond to a range of wave lengths. We could put down a scale and specify a wave length λ (the Greek letter lambda) at each point. This is one important aspect of a spectroscope—a specification of just how wave lengths are spread out by the instrument. A second concerns the size of the image of the slit using light of a single wave length. As we said, there is a smallest possible size, and this corresponds to a certain wave-length interval on the wave-length scale. Let us call this wave-length interval $\Delta\lambda$ (Δ is the Greek letter capital delta). It is a critical number in spectroscopy. It is the smallest "color" difference in the spectral range that the instrument under discussion can separate or resolve. In the world of radio it would correspond to the smallest frequency difference between two stations that a receiver can tune in on separately. Scientists find dimensionless numbers to be especially useful, since one does not have to know the units used in order to understand the significance of a quantity. They describe the ability of an instrument to separate wave lengths not in cm, or millimicrons, but in terms of a dimensionless resolving power. If λ is the wave length we are operating at, and $\Delta\lambda$ is the range of wave lengths occupied by the smallest slit image we can produce, then the resolving power is measured by the pure number

$$\frac{\lambda}{\Delta\lambda}$$

A Cadillac of a spectroscope would have a resolving power of 10^5 to 10^6, or a hundred thousand to a million. This power, while exceedingly good, is not yet in the specially designed, exquisitely adjusted racing car class, but well beyond the Fig. 5 type of instrument.

With a good spectroscope, sunlight can be broken up to show the pattern of Fig. 6. Here we have all

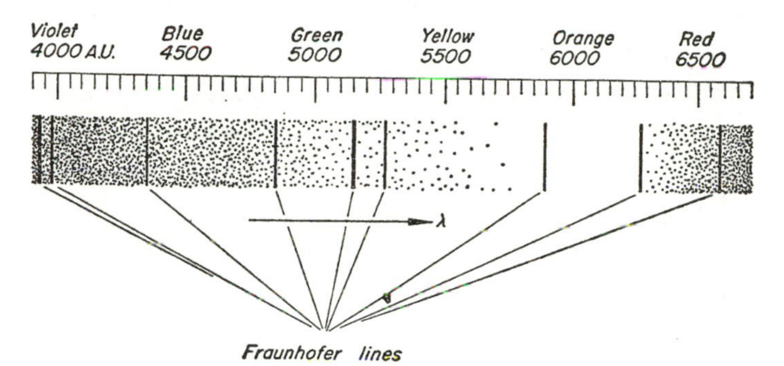

Fig. 6. The spectrum of the sun has the full range of colors, beginning with red at the right. The black Fraunhofer lines are also visible. A.U. is the angstrom unit, which is 10^{-8} cm.

the colors of the rainbow. The actual spectrum of the sun covers a much greater range, out beyond the red into the infrared and then the microwave and radio range, and out beyond the blue and violet into the ultraviolet, and then the X-ray and gamma-ray range. The continuous spectrum is crossed, however, by dark lines, certain wave lengths that are present in much less intensity than the rest.

These missing wave lengths in the spectrum of the sun point the way to the next step in our train of thought. The light beam can carry messages about reality beyond our senses. We must use instruments to detect them. There is something sharply tuned involved in these narrow Fraunhofer dark lines, as they are called.

The reader will now wonder what on earth "Fraunhofer" lines are. Dark lines in the spectrum of the sun may be clear enough, but what has this to do with something called "Fraunhofer"? Physicists, like people in general, often name things for people associated with their discovery, or development. We are not put out by hearing of Marie Byrd Land, or the Hudson River, or Pike's Peak. It might be of interest to know a little of the personality and character of the people involved, but this would hardly be relevant to the geographical features which they designate. Similarly, we shall run into a lot of names in physics. The volt, ampere, and ohm mean something to most readers as physical quantities, apart from the people for whom they are named. Similarly the "Fraunhofer" lines are usually referred to by name, rather than as "the dark lines in the solar spectrum." We shall run into a variety of such new names, the Balmer series, the Rydberg constant, and so on. There is no help for it but to learn what they mean.

Let us then get back to our theme concerning the solar spectrum and the Fraunhofer lines. The sun is a great jangling mass of particles that hit each other so violently that only a great optical "din" of all possible wave lengths results—namely, the almost continuous spectrum of Fig. 6. But outside the sun are cooler clouds of atoms, and these absorb their natural frequencies, just as a piano string will absorb sound of its own frequency, as any reader may verify for himself by singing into a piano with the pedal down. The dark lines are frequencies characteristic of certain kinds of atoms. We can match them exactly with the light from lamps here on earth.

The Hydrogen Spectrum

If we wish to look more deeply still into light, we find ourselves looking at the radiating atom. Let us take the simplest of all atoms, the hydrogen atom, made up of a positively charged heavy nucleus, with a single electron orbiting around it, and let us strike it just hard enough to make it vibrate electrically, and send out light. What sort of optical chords will it radiate? It is usually the simplest patterns that are recognizable as such in our human way of looking at things. The patterns in the sunlight are complex and hidden. In the hydrogen atom they are more easily seen.

A spectrum of the light from hydrogen atoms is shown in Fig. 7, and a pattern is immediately discernible. We recognize, by using a spectroscope, that the pinkish light emitted from a lamp containing atomic hydrogen is made up of a series of wave lengths, or lines as they appear in the image produced by the spectroscope, and that there is a regularity about their arrangement. At this point the second comment about "seeing" things is applicable. We need not only an instrument, but numbers. If there is some physical significance in this pattern, it must be

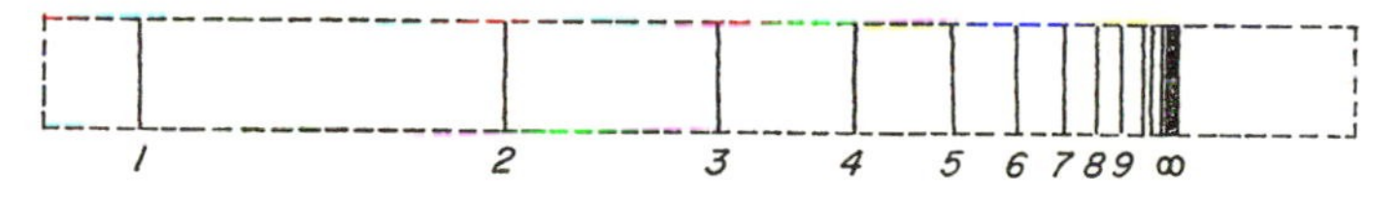

Fig. 7. The series of lines called the Balmer series are found in the light emitted by hydrogen atoms.

related to the wave lengths, or to the frequencies of the radiations emitted.

And here the real significance of physical measurement begins to appear. If we can arrange a series of facts in some significant way, or in some mathematically expressible form, then there is hope that this phenomenon can be incorporated into the "laws" of physics, into the body of knowledge that we feel we understand because we are aware of many complex interrelations. We have a puzzle whose solution we must find by trial and error. We must "play" with numbers.

To get back to the hydrogen spectrum, here are the numbers describing it, starting with the long wave lengths on the red side of the spectrum, and proceeding toward the violet, where they are more and more closely spaced.

TABLE I

WAVE LENGTHS OF THE LINES IN THE BALMER SERIES OF THE SPECTRUM OF ATOMIC HYDROGEN

Line Number	*Observed Wave Length As Listed by Balmer*	*Computed Numbers Using Balmer's Formula*
1	6562.10×10^{-8} cm	6562.08×10^{-8} cm
2	4860.74 " " "	4860.80 " " "
3	4340.10 " " "	4340.0 " " "
4	4101.2 " " "	4101.3 " " "
5	3968.1 " " "	3969.7 " " "
6	3887.5 " " "	3888.6 " " "
7	3834.0 " " "	3835.0 " " "
8	3795.0 " " "	3797.5 " " "
9	3767.5 " " "	3770.2 " " "

Solving this puzzle is about as interesting as betting on a horse race that already has been run. The real interest is in the outcome. But the action as seen in a newsreel has its points too, and that is all that is left for us at this point.

The Significance of Physical Measurement

The events with which we are concerned cover about a twenty-five-year period. Accurate detailed information on atomic spectra were accumulating in the 1860s. The first observations concerned the similarity of spacing of groups of lines, like closely spaced pairs. From 1870–85 a half dozen discussions were published, but all seemed to reach a dead end after an initial striking success, like the observation by George Stoney, published by the London, Edinburgh and Dublin *Philosophical Magazine* in 1871, that the number 131,277.14 divided by 20, or 27, or 32 led to the following, which seemed to duplicate three wave lengths in the hydrogen spectrum:

Observed Wave Lengths Listed by Stoney	*Calculated Numbers*
6563.96×10^{-8} cm	6563.86×10^{-8} cm
4862.11 " " "	4862.12 " " "
4102.37 " " "	4102.41 " " "

However striking a partial success like this may be, science is likely to forget it unless it leads somewhere.

And now observe the difference between the above "failure" and a more lasting "success." Johann Balmer, in 1885, showed that the following formula gave the numbers listed as "computed" in Table I:

$$\lambda = 3645.6 \times 10^{-8} \frac{n^2}{n^2 - 4} \text{ cm}$$

where n takes on the values 3, 4, 5, 6, and so on. This use of integers is vaguely reminiscent of overtones in musical instruments. The relationship between the frequencies of the fundamental tone and the overtones of strings, or organ pipes, similarly makes use of integers. Stoney and the other analysts were all looking for some kind of relationship using simple integers.

The agreement shown in Table I is most startling and mysterious, at least until it is explained, and it illustrates beyond equivocation the great simplicity—one might even say beauty—in this hidden world beyond the range of our five senses, but available to us if we use our ability to think, to build aids to our senses, and to present what we find in a suitably digested form. And if a procedure, or formula, not only describes a body of facts, but illuminates other areas, then it is truly revered and cherished. As we shall see, Balmer's formula is really a gateway, a pass in the mountains, by which mankind entered the world of atomic physics. The formula for the Balmer series can be rewritten

$$\frac{1}{\lambda} = \frac{4 \times 10^8}{3645.6}\left(\frac{1}{2^2} - \frac{1}{n^2}\right) \text{cm}^{-1}$$

or

$$\frac{1}{\lambda} = R\left(\frac{1}{l^2} - \frac{1}{n^2}\right)$$

$$R = 109{,}730 \text{ cm}^{-1}$$

$$\text{and } l = 2, \quad n = 3, 4, 5, \ldots$$

There are other radiations from the hydrogen atom, in the ultraviolet, and in the infrared. The wave lengths radiated can *all* be represented by the above formula with

$$l = 1, 2, 3, \ldots$$
$$\text{and } n > l$$

This is called the Rydberg formula and is described in the same *Phil. Mag.*, as it is colloquially known, for the year 1890. The constant R, seemingly arbitrary, is called the Rydberg constant in J. R. Rydberg's honor.

Our inner eyes are almost peering at the hydrogen atom. We have detected the signals, but we don't yet know what they mean. We have found a pattern, but is that all that is in store for us mortals? Can we hope to understand how such patterns come about? In a sense, *yes.* We can find simple "laws" or "principles" from which such patterns follow. But also, in a sense, *no,* because we don't understand where the laws come from.

And so, as scientists, we must be prepared to take the next step. Is the Rydberg formula really correct? We must make more accurate measurements and compare the results to theory. And when we examine these spectral lines from hydrogen, which seem to be represented so accurately by a single simple formula, we find that they are not single lines, but appear to have a structure.

For instance, the line number 1 in the Balmer series, when examined with a spectroscope of very high resolving power (actually a grating), yields the result shown in Fig. 8. The Rydberg formula isn't exact at all! Should we discard it and start all over? Certainly not. Physics is full of such situations. We find a relationship which seems to be true to a certain extent under certain conditions, and in our enthusiasm conclude that it must be absolutely true always. This is a most dangerous pitfall. "The surface of a pond seems flat. Therefore the earth is flat." Wrong, of course. "We can describe the motion of objects like bullets, pendulums, billiard balls, etc., very accurately assuming that their masses are constant. Therefore, the masses of all objects are constant."

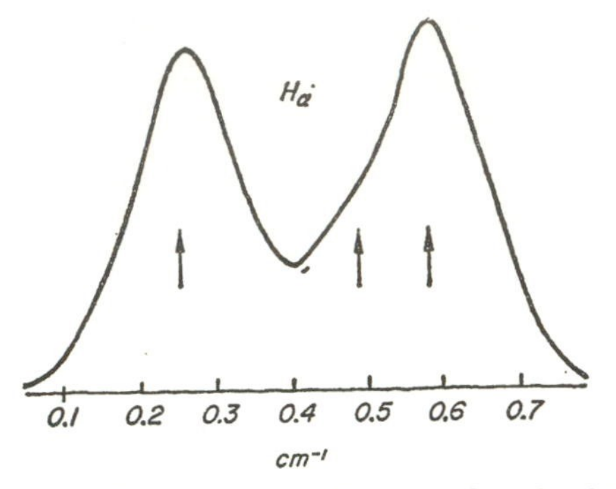

Fig. 8. The actual structure of line number 1 of the Balmer series as revealed by a high resolving power spectrograph shows three components. Two of the components are at the peak points, the third at the point of the curve indicated by the middle arrow.

Wrong again. As objects move faster and faster, much faster than we usually see them, their masses increase. This doesn't mean that road maps of a flat earth should be discarded, or that Newtonian mechanics should be scrapped. But when we are dealing with travel over large segments of the earth, or motions approaching the velocity of light, we must use appropriate revisions. Similarly, Rydberg's formula is perfectly valid to a certain approximation, but it is not exact. Even hydrogen atoms are not that simple.

But now we must wind up our train of thought, although it really has no end. There appears to be more and more to see, the deeper we look. To drive this home, we take one more step–into the hydrogen atom. What we see there does not leave us so much a humanly satisfying picture, as the power to predict, and this is of course the really exciting aspect of our knowledge. We can watch the collision of two billiard balls, but this is not as significant as our ability to predict. We have a method, a theory, really a black box into which we can feed the relevant information.

Billiard ball *A*, weight so and so, speed so and so, position so and so, is moving in direction so and so with such and such speed. Billiard ball *B*—comparable data. Then out of the black box comes the answer —collision to take place at point *P* at time *T*. Velocities of *A* and *B* after collision will be so and so. Everyone will grant that this can be done and will understand to some extent how. Any billiard player can do it in his head. If an electron were a billiard ball, this same black box would do. But it isn't. We mustn't carry forward normally acquired experience into a region beyond our senses without being very careful. And after many mistakes we have learned how to build a black box for electrons. We feed into it very little information. For the solution of the hydrogen atom problem we feed in only a few facts. The mass m and charge e of the electron. The fact that the electron is attracted to the nucleus according to an inverse square law of force. And finally a constant, called Planck's constant h, and the velocity of light c, both of which appear over and over in physics. These constants are all known from independent measurements. They have the values, according to current handbooks,

$$m = 9.1085 \pm .0003 \times 10^{-28} \text{ gm}$$
$$e = 4.80281 \pm .00008 \times 10^{-10} \text{ esu}$$
$$h = 6.6251 \pm .0002 \times 10^{-27} \text{ erg sec}$$
$$c = 2.997928 \pm .000004 \times 10^{10} \text{ cm/sec}$$

The little black box takes in this information and comes out with the answer for the wave lengths radiated by the hydrogen atom.

The answer the box gives is

$$\frac{1}{\lambda} = \frac{2\pi^2 me^4}{ch^3}\left(\frac{1}{l^2} - \frac{1}{n^2}\right)$$

where l and n are positive integers greater than zero,

and n must be greater than l. This is, of course, the Rydberg formula, with the additional information that the constant R has the value

$$R = \frac{2\pi^2 me^4}{ch^3}$$

If we substitute the values for these constants, we find for R the number, 109,700. Whether this degree of agreement with experimental data is really satisfactory is another subject, which we will not take up. For our purposes it is excellent. We clearly have a powerful black box for describing the behavior of electrons, and we can use it in many ways.

And so on. The inclusion of the magnetic effects due to the electron leads to further agreement, but only partial agreement, with the structure of the lines, as illustrated in Fig. 8. We get better agreement when we consider that the hydrogen atom is not just in empty space, but that there are all kinds of electromagnetic fluctuations and waves in this space, and that they too are influencing the atom. So we have signals not only about the electron plowing its way around the nucleus, but about what goes on in "empty space"! And then we find that there are even finer effects which we can account for in a further look into the atom if we take account of the fact that not only the electron is a little magnet, but also the nucleus. We notice that the wiggling of the magnet in the nucleus is affected in ways that we can observe in great detail by studying the atoms around or near the one that we are observing. We have then a way for looking into solid matter as well as atoms!

We have come a long way in our train of thought —from the village, to the church, to the skeleton, to the light, to the atom, and on. It is not fashionable to speak of scientific matters in a philosophical or religious context. But I am continually struck by trains of thought like the one I have sketched out. The world we live in is clearly not a random jumble of objects, forces, and motions. There is a design. There is every indication of a creation, of complex ordering tendencies. Are not the patterns which lie beyond what men or beasts or insects have made, the patterns that we can usually see only with our inner eye, are they not real clues to the nature of creation, and the Creator? And is it not conceivable that you who read this, and your children, and your children's children, will find new and deeper meanings in these clues?

CHAPTER 2

Data

This chapter is intended to introduce the reader to the peculiar fascination of collecting and examining facts, the kinds of facts that are called the *data* of science. At first glance the man who peers for long, long hours through a telescope at the stars, who gets stiff and cold and often discouraged trying to get a few better observations than actual circumstances at the moment permit, such a man will seem a "peculiar" sort of fellow. In general, the kinds of observations that a physicist makes, measuring little marks on a photographic plate, or watching the pointer or the spot of light on a scale controlled by an electric meter, or, nowadays, the pages and pages of numbers typed out for him by a machine, all these seem abstruse and forbidding to the casual eye.

To tell the truth, there are times when this or any kind of work seems dull, exhausting, even fruitless. When, after days of trying, you still can't find the leak in the vacuum system, or when, after you patiently have fitted an elaborate piece of apparatus together, an oscilloscope suddenly picks up a lot of meaningless "noise" from some unknown source, or when you have been working all day and all night on a series of measurements that must for some reason be completed at once, then of course the work is unattractive. This is no different from the difficulties and frustrations to be expected in any serious undertaking. The question to be answered does not concern the nature of such discouragements, but rather why any sane man would choose to spend his time in the pursuit of experimental physics, even when everything is going just right, his apparatus performing as planned, and he can do his work seated in a comfortable chair.

Fact Gathering

To answer this, let us forget physics for a while, and consider the kind of fact gathering that any reader would agree made sense. The key is in the usefulness of the facts. A man who spent his life counting the fence posts along country roads we might dismiss as an idiot, or at best simple-minded, especially if he himself had no better explanation than that he liked counting posts. But a group of men measuring distances and angles along the streets of Chicago, for example, are in a quite different category. They are paid for their "work," because the data they collect are useful for making maps and fixing property boundaries. A stranger who wants to find his way around Chicago must have a map, and the buyer of land must have a description, and this must be based on reliable facts. Another set of facts that would generally be accepted as interesting would be the ingredients and proportions of a loaf of bread, or the time and temperature required to roast a chicken. Even though these facts may not nowadays be needed by a housewife who will settle for bread and ready-to-serve roasted chickens from the supermarket, the facts are still needed by someone as long as there is a demand for bread and roast chicken.

But these are obvious facts which every reader can multiply almost endlessly with examples from his daily life. Though useful, they are usually not very generally interesting. A more interesting type of fact leads more toward our understanding of a situation than toward direct use. Consider the facts that have led us to believe that the earth is round, rather than flat, and that it is one of several planets moving around the sun. These facts are of course "useful" in connection with space travel, for instance, or navigation, but quite apart from this, most people must have a deep satisfaction in understanding why you must change the setting of your clock as you travel around the surface of the earth, or why from time to time the sun and moon are eclipsed.

Various kinds of understanding are based on various kinds of facts. Some we can interpret and use immediately. A glance at a human face or at a dog standing in front of us provides data that our brain can analyze in less than a second, and produce in us an understanding of a situation on which our decisions for action depend. Facts about human illness and health require a long time to be collected and examined to the point at which we understand the underlying causes and can expect to find reliable therapeutic measures. Facts about the universe around us have been gathered and examined as long as there have been creatures with senses and minds with which to notice and remember.

And now, as we take up the world around us, a new element enters into our collecting of data—the unexpected. Or perhaps we had better say, the expected unexpected—like catching a fish. We no longer measure and record only in order to satisfy a need. We don't know what we shall find. Some strange instinct guides us in deciding what observations to make and record because we hope, at some later time, to find an interesting, revealing, or useful pattern in the collected and analyzed facts. This is the instinct that has led men to observe and record the movements of the stars and planets, of the sun and moon, year by year more carefully, until Johannes Kepler found in Tycho Brahe's papers enough data about the directions in which the planets were observed at various times to conclude, after discarding many other hypotheses, that they really were moving in elliptical orbits around the sun, with the sun not in the center, but at one focus, and this in turn was a sufficient analysis of the raw facts to allow Isaac Newton to conclude that a gravitational force determined not only the motions of projectiles and falling bodies on the earth, but of the entire solar system.

Raw Data

Taking down data is a very personal thing. As an arbitrary example of how the variation of one factor influences a second, let us consider the following electrical problem with which some readers may be familiar. Suppose one wants to determine how the current through a low-pressure gas discharge like a fluorescent lamp varies as the applied voltage is changed. This can be done using a setup like that shown in Fig. 9a. An ammeter is connected in series with the lamp to read the current through it. A voltmeter is connected across it to measure the voltage required to drive the current through the lamp. Of course, care must be taken to make the resistance of the voltmeter sufficiently high so that it does not drain away from the lamp an appreciable part of the current measured by the ammeter.

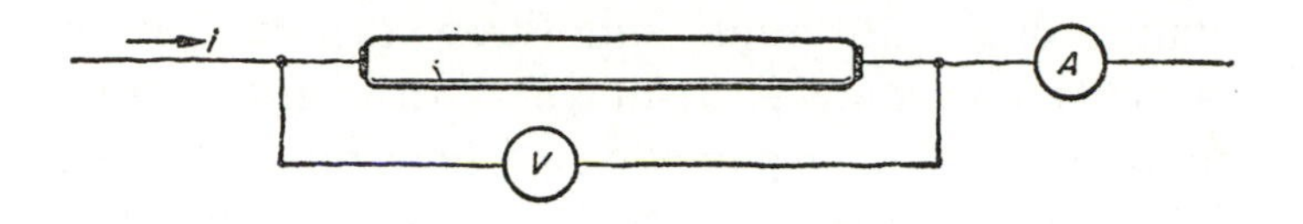

(a)

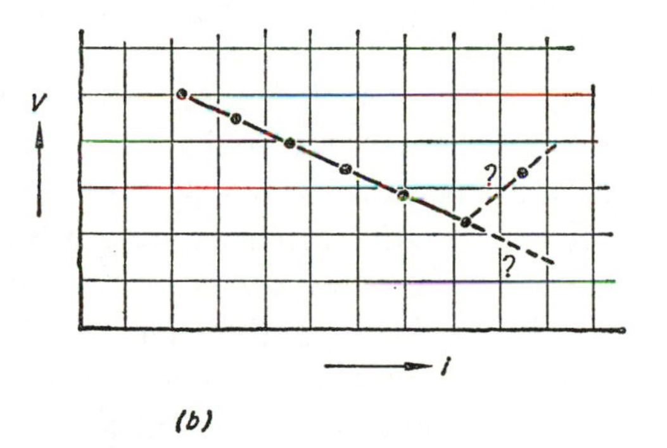

(b)

Fig. 9.

Some experimenters enter their findings in a notebook, or merely scribble on the back of an envelope two lists of numbers, for current and voltage. I, myself, usually plot it as I go along, entering in our example the observed voltage for any given current, so that I can see whether the points lie on a smooth curve, whether there are any dubious points like the last point in Fig. 9b perhaps due to a mistake in reading, or whether there are some interesting irregularities such as a kink in the curve that should be investigated further before the setup is changed. Clearly more readings are needed to settle this. Then, when the measurements are completed, what do you do with the results? Let us suppose that they are to be part of a report, or publication, and that in the final version a graph showing the experimental observations will be most suitable. Some people might simply draw the graph and throw the data away. Another procedure is to make a permanent record, together with an explanation of how the data were taken, and how reliable they are, and to include a discussion of the observer's interpretation of their meaning. Especially in industrial research, where patents may be involved, it is of the greatest importance to keep records.

I myself have not kept such records. Although I have a file containing many notebooks, they are full of messy scrawls that became meaningless, or at least difficult to interpret, shortly after they were entered. I might be able to reconstruct how published curves were arrived at because I might remember details that were not entered. But such notebooks would be very poor evidence in court. On the other hand, I have the greatest admiration for the achievements and records of John Waymouth, a friend who works for the Sylvania Engineering Laboratories. He usually takes data on a scrap of paper. We have often discussed these data as they appeared, hour by hour, day by day, week by week. But unless they are obviously meaningless, perhaps due to the use of a faulty instrument, or neglect to control some vital factor in an experiment, he enters in his notebook a discussion, data, graphs, everything relevant. Sample pages of data, and of the index kept at the back of each book, are shown in Fig. 10. His lab is usually in a terrible mess. Keeping his desk clean is of no interest. His mind is on an experiment, or a calculation, not on distracting details. But behind his desk is a bookshelf containing a good many reference books, and his notebooks. When in the course of discussions, an old train of thought is retraced, but new questions arise, he can reach out for the one or two books that recorded the previous work, look in his meticulously prepared index for the page where the experiment in question was described, and usually in a few seconds cut through a haze of assumption and argument with reliable, known facts.

The facts, or data, that we shall deal with here are of a quite different form. They are contained in Table II (see Appendix). We propose to use these data to establish certain properties of the motions of the earth and some of the planets. The table contains many more items of interest that the reader may work out for himself. The listed numbers are clearly not direct observations. They purportedly relate to observations at regular ten-day intervals starting from September 14, 1954, and extending right past the date of writing of this account, in July 1962, on to February 7, 1971. Today, in order to specify the position of a planet in the sky, we do not have to go to an observatory for information. We know these motions well enough to calculate them accurately. The table, then, is produced by calculation rather than observation, but it is completely consistent with all observations. It was, in fact, prepared by Dr. Owen

6048- 68

Sylvania Electric

Title or Project: ZF 1081 original sample ; original erasure

Techniques: See 3457-29, 2740-2, 14

time	action	voltage	shunt	do	dm	sd.
0	irradiate	0				
30	stop	0				
120	charge	+1000	1.0	-6		0.00
180	discharge	0	1.0	-6.65	-1.94	4.71
180	charge	1000	1.0	-7.09	-6.33	0.76
190	discharge	0	1.0	-7.10	-3.08	4.02
0	irradiate	0				
30	stop	0				
120	charge	1000	1.0	-6.85	-6.85	0.00
150	discharge	0	1.0	-7.25	-3.90	3.35
180	charge	1000	1.0	-7.50	-6.15	1.35
210	discharge	0	1.0	-7.65	-3.80	3.85
0	irradiate	0				
30	stop	0				
120	charge	1000	1.0	-6.80	-6.80	0.00
220	discharge	0	1.0	-7.50	-5.75	1.75
250	charge	1000	1.0	-7.70	-5.65	2.05
380	discharge	0	1.0	-7.55	-4.52	3.03
0	irradiate	0				
30	stop	0				
120	charge	1000	1.0	-6.70	-6.70	0.00
420	discharge	0	1.0	-7.75	-6.70	1.05
450	charge	1000	1.0	-7.65	-4.75	2.90
480	discharge	0	1.0	-7.60	-4.80	2.80
0	irradiate	0				
30	stop	0				
120	charge	1000	1.0	-6.70	-6.70	0.00
720	discharge	0	1.0	-7.53	-6.89	0.64
750	charge	1000	1.0	-7.50	-4.25	3.25
780	discharge	0	1.0	-7.46	-5.08	2.38

Witnesses Date:

Signature: John D. Waymouth Date: Dec. 28 '53

Fig. 10.

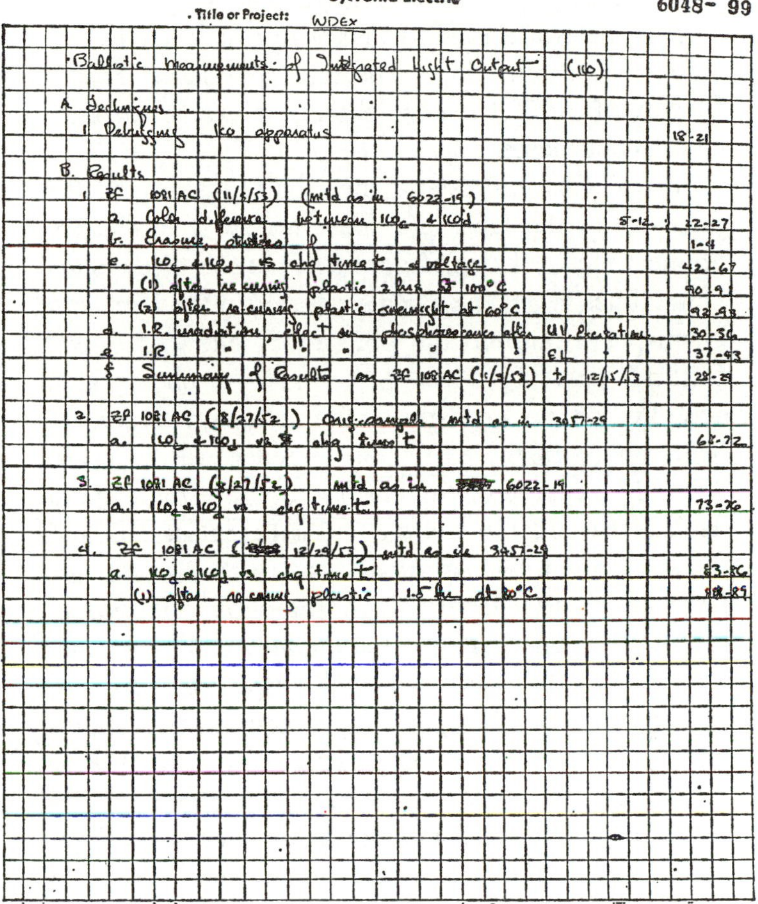

Sylvania Electric

6048- 99

Title or Project: WDEX

Ballistic measurements of Integrated Light Output (ILO)

A. Techniques

1. Debugging ILO apparatus — 18-21

B. Results

1. ZF 1081 AC (11/3/53) (mtd as in 6022-19)
   - a. Color difference between ILO_c & ILO_d — 5-12; 22-27
   - b. Erasure studies — 1-4
   - c. ILO_c & ILO_d vs chg time t & voltage — 42-67
     - (1) after re-curing plastic 2 hrs at 100°C — 90-91
     - (2) after re-curing plastic overnight at 60°C — 92-93
   - d. I.R. irradiation, effect on phosphorescence after UV excitation — 30-36
   - e. I.R. " " " " EL — 37-43
   - f. Summary of Results on ZF 1081 AC (11/3/53) to 12/15/53 — 28-29

2. ZF 1081 AC (8/27/52) orig. sample mtd as in 3057-29
   - a. ILO_c & ILO_d vs chg time t — 68-72

3. ZF 1081 AC (8/27/52) mtd as in 6022-19
   - a. ILO_c & ILO_d vs chg time t — 73-76

4. ZF 1081 AC (12/29/53) mtd as in 3057-29
   - a. ILO_c & ILO_d vs chg time t — 83-86
     - (1) after re-curing plastic 1.5 hr at 80°C — 88-89

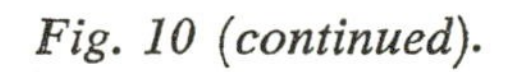

Fig. 10 (continued).

Gingerich, of the Smithsonian Astrophysical Observatory, for historical purposes. The primary purpose was to specify the positions of the planets in the past, so that historians might quickly find just where any planet was at any time. The table published here is by way of a slight readers' bonus, of value for educational purposes.

In order to understand the significance of all the numbers in the table, we must review a few astronomical facts. The chief apparent motion of the stars is in circles about the earth's axis, or in the Northern Hemisphere, around the polestar at one end of the Little Dipper. In Plate I we have a photograph, actually a time exposure, obtained with a camera held rigidly on the earth and aimed at the polestar. Actually, because of the earth's rotation, the camera itself is traveling on a circle and is being rotated, and the streaks in the illustration result from this rotation. By rotating the camera in a direction opposite to the rotation of the earth, the camera will essentially be held in a fixed position with respect to the fixed stars, and they will appear as points on the photograph, even with a time exposure, as in Plate II. But, as you can see from this photograph, there are exceptions. Among the great mass of fixed stars there are a few "wanderers," that move with respect to the others. These are the planets, asteroids, and comets, and, of course, the sun and moon. We shall attempt to deduce the actual motions of some of these in space from their apparent motions among the fixed stars.

The Path of Scientific Investigation

In any scientific investigations, or in any serious attempt to understand physical data, there usually are certain common elements. We shall illustrate some of these in the following pages. These crucial steps are to a great degree responsible for the pleasure and excitement of taking down data and studying them. Try to notice some of these as we go along. The main one is inspiration, or motive. In our case it is rather flat, because the answer given by the data is already accepted by everyone–namely, that the sun is at the center of a rotating solar system. In order to present the facts in their historical perspective, we shall start with the assumption of an earth-centered system. First try to put yourself into a frame of mind to accept this, and as one bit of evidence after another falls into place, let yourself realize little by little the great beauty and relative simplicity of the heliocentric system. A second crucial step is the choice of the particular way you look at the facts. Here, too, inspiration is needed, and also experience. It is like a general's choosing a battleground on which his forces can maneuver and fight to advantage. In our case it is the choice of a coordinate system. There are usually many complicated and confusing ways of looking at a set of facts, but primarily one or two that will show the essentials elegantly and convincingly. And finally, in spite of one's best skill and judgment, complications arise with which one doesn't know how to cope. Often they are mathematical problems that one is convinced have a solution but that involve ideas and relationships beyond the experience of the experimenter. A certain amount of mathematical experience is indispensable. A research physicist (not to say every educated person!) should know geometry, algebra, how to differentiate and integrate, and especially how to use and where to find textbooks, handbooks, and tables. And, in addition, he should know a few experts in other fields who are willing to help, especially by pointing the way in which help can be expected.

In order to get started, let us look at the rough facts about the motions of the wanderers. How *do* they move? If it is the motion with respect to the fixed stars in which we are interested, we must begin with a star map. The stars appear to be on a sphere surrounding the earth. A star map centered around the polestar is shown in Fig. 11. A spherical

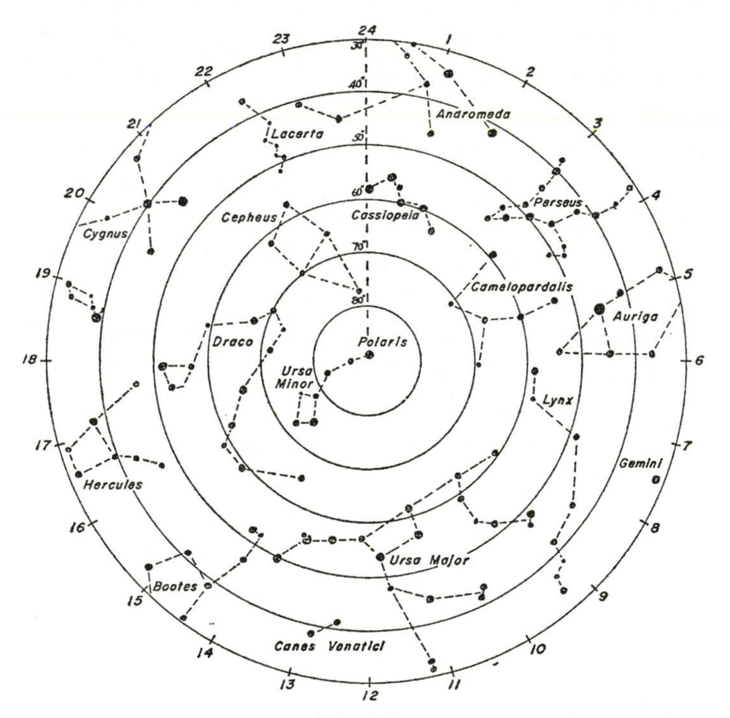

Fig. 11.

distribution is notoriously difficult to represent accurately on a flat piece of paper. A common form of projection, often used for world maps, is the Mercator projection, in which a band around the earth's equator is apparently peeled off and flattened out so that north is always straight up on the final map. This projection produces a fairly accurate picture of geographical relationships near the equator, but greatly exaggerates the east-west dimensions near the poles. A star map made in the same way, and with the same north-south axis as the earth, is shown in Fig. 12.† The map shows only the equatorial region, and the vicinities of the poles are missing. Constellations, and the usual designations for many stars, are given. On this background is a dashed line along which the sun and planets move.

Here the reader should pause to examine the chart with a little care. We are going to make some judgments about how to proceed, and the first step is to look at the facts and think about them. Does it not seem strange that the position of the sun can be plotted accurately among the stars? The sun is visible only in the daytime when the stars are invisible. It is all very well to toss this off by saying that this is all calculated data anyway, but how do we get relevant information to feed into the calculating machine? Just think about it. One way of thinking about it is to consider where, with respect to us, the sun is halfway between sunrise and sunset. Then, where is it halfway between sunset and sunrise? If we point to this same place in the sky on the previous night

† This and other star maps are reproduced from maps published by the Sky Publishing Company, 51 Bay State Road, Cambridge 38, Massachusetts, and may be obtained from them.

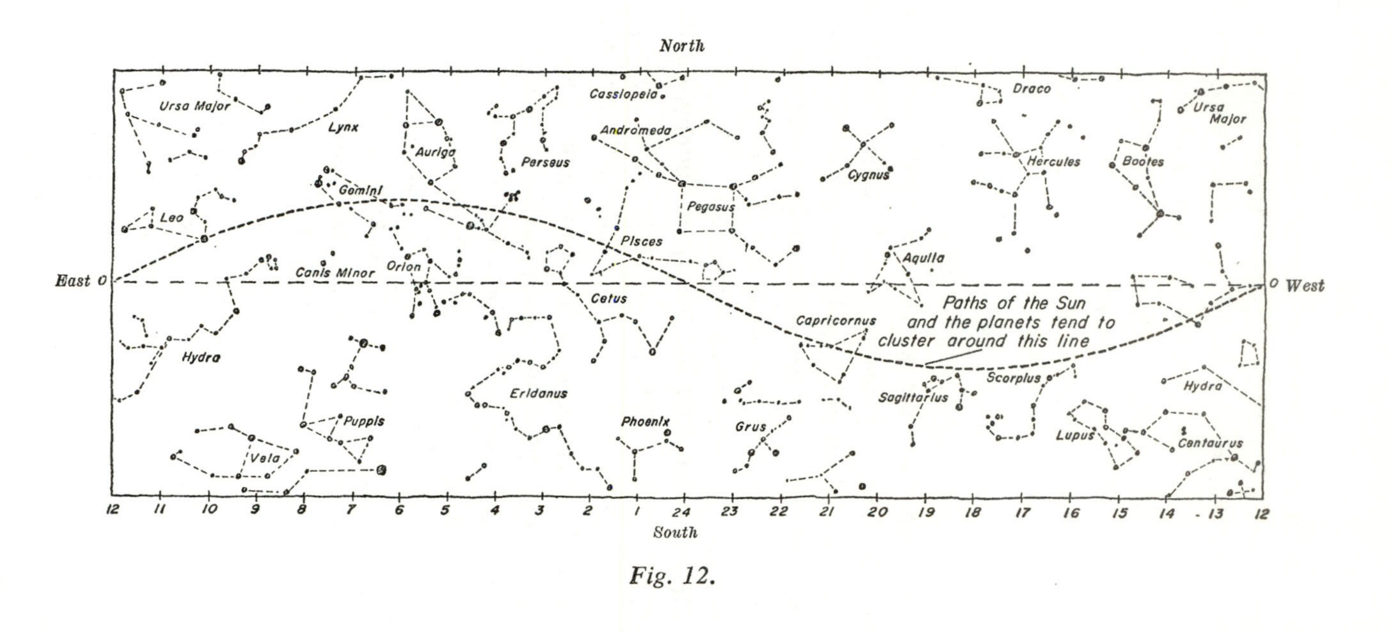
North
East 0
0 West
South
Ursa Major
Lynx
Auriga
Perseus
Cassiopeia
Andromeda
Cygnus
Draco
Hercules
Bootes
Leo
Gemini
Pegasus
Pisces
Aquila
Canis Minor
Orion
Cetus
Capricornus
Paths of the Sun
and the planets tend to
cluster around this line
Hydra
Eridanus
Phoenix
Grus
Sagittarius
Scorpius
Lupus
Centaurus
Puppis
Vela
12 11 10 9 8 7 6 5 4 3 2 1 24 23 22 21 20 19 18 17 16 15 14 13 12

Fig. 12.

and on the following night (and at night we can locate this with respect to the stars if the sky is clear), can we not conclude that the sun is in an exactly opposite direction, and so locate it on the star map? Then we might think about the part of the sky on the map which is seen at night, and the part which is invisible because of the overpowering daylight. Now it is clear that this depends on our position on the earth. If we are on the equator, the zero latitude line will pass directly overhead, and the part of the sky visible to us in the dark will be roughly more than 90° from the sun. Why "roughly"? Well, when the sun is on the celestial equator, or at the position of 0° or 180° longitude, day and night will have equal length. But what about conditions when the sun is not on the celestial equator? This is an interesting problem, but beside the point for our discussion. In general, it is clear that planets or stars will not be visible when they are near the sun, and data on such stars or planets are inferred or calculated.

To continue our considerations of the facts about the motions of the planets, the next obvious question is why the motion is along the curved path shown, rather than some straight path, which might at first seem more "natural," and here a first inspiration, or what was originally an inspiration—a good idea—comes in. If the earth is not at the center of the solar system, should we not perhaps consider a different description of the star positions? We have chosen the North Pole simply because the earth is spinning on an axis passing nearly through the North Star. How would matters look if the sun were not moving around the earth, but the earth around the sun? Would it not be sensible to consider a Mercator projection whose polar line was the axis on which the earth moves around the sun? These relationships are shown in Fig. 13. The figure is drawn with the axis of revolution of the earth around the sun making an angle of about 23° with respect to the axis of rotation of the earth. In these circumstances what would be the apparent motion of the sun among the stars?

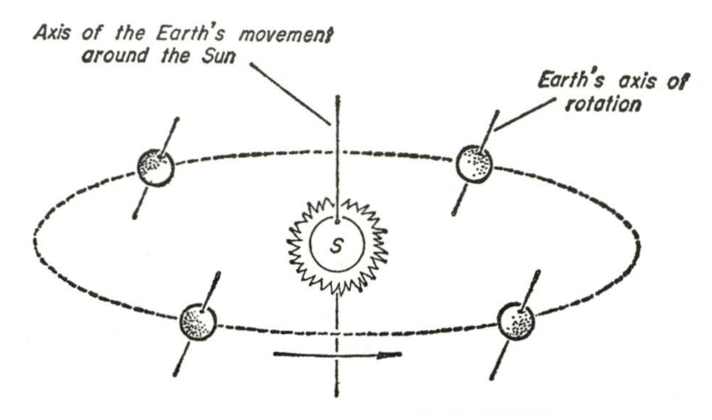

Fig. 13. The direction of the Earth's axis of rotation remains fixed as the Earth revolves around the sun.

The motion of the earth around the sun defines a plane with respect to the fixed stars in space, and the apparent motion of the sun as seen from the earth would be defined by the intersection of this plane on the map of stars. We have sketched in the star map as it might be wrapped around the earth in Fig. 14. The intersection which we require is an oval, partly above the equator, and partly below. If we open up this map, as in Fig. 12, we see that the path of the sun is qualitatively that found from our construction.

And here we come to a point where we should like to convert a qualitative observation, like the

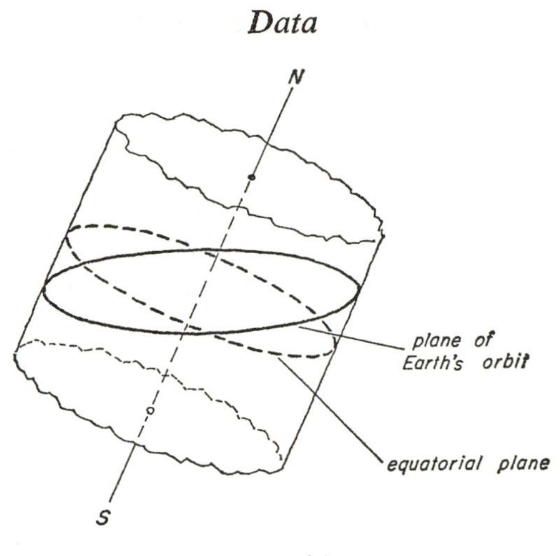

Fig. 14.

above, into a quantitative one, to test whether the apparent motion of the sun through the stars is *exactly* that which would be expected if the sun moved in an orbit in a plane. This sort of thing can be an awful headache, especially if you fumble around clumsily, making approximations and mistakes, getting confused over bad notations, and so forth. This is not the place actually to present the proofs, but at least we can have a look down this vista which in one form or another will open up repeatedly before the scientist. How shall we specify the position of a star? One obvious way is by longitude and latitude, as we specify the position of a point on the globe. In Fig. 15a we show two angles, ϕ and θ, describing the position of a point P. How do we specify *exactly* the latitude and longitude of every point on a great circle whose axis makes an angle with the axis of our coordinate system? Or, in order to solve this problem, would it not perhaps be useful to find out what the latitude θ' and longitude ϕ' of the point P are in some other coordinate system, as in Fig. 15b? Then

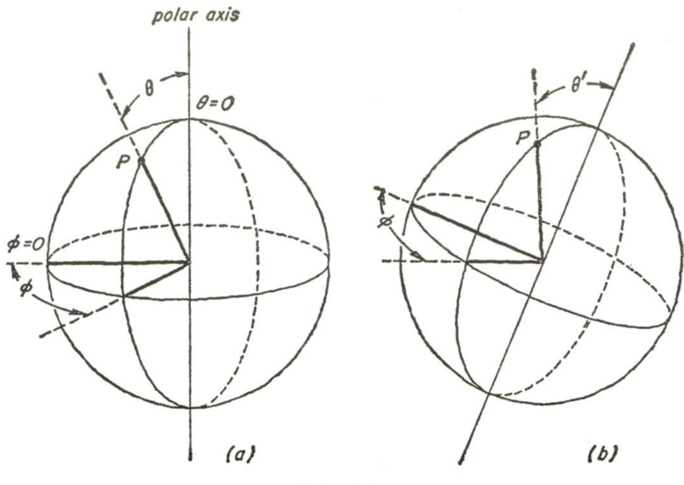

Fig. 15.

we could specify that a great circle can always be represented in some coordinate system—namely, that one with its axis perpendicular to the plane of the great circle—as the locus of points for which $\theta' = 90°$ and ϕ' goes from 0 to 360°. Is this a dull procedure? Or an exciting one? It all depends on you. If you are interested in finding out, as a result of direct observation, whether the sun really appears to move in a plane with respect to the fixed stars, or whether there is for some reason a bump or irregularity (astronomers would call it a "perturbation") in its path indicating some new and unexpected disturbing force, it may be exciting.

Let us suppose that all this has been done, and that we have calculated from Fig. 15 how to draw the same star map in a new coordinate system in which the plane of the earth's orbit defines the equator. The paths of the planets follow this same equatorial line very closely. Now the star map and the plane in which the planets move are shown redrawn in Fig. 16.

This representation is, perhaps, a good one for our basic raw data, simply the angle along this equatorial plane in which we may see a planet at any time. This angle is the basis for the designations in Table II, which you will find in the Appendix and from which we print here two short excerpts. We choose these particular excerpts not only because they illustrate the

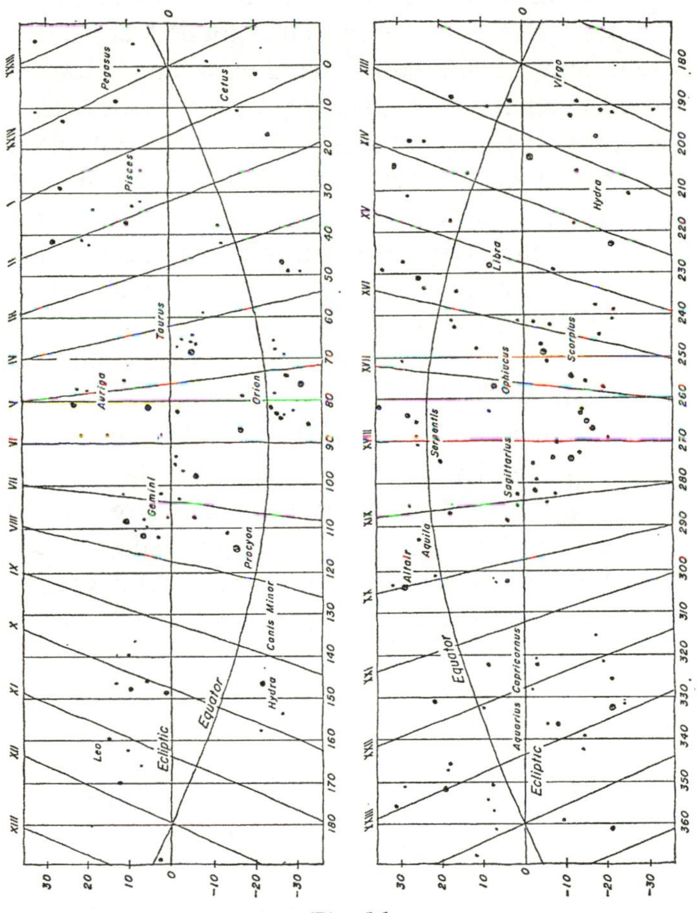

Fig. 16.

pattern of the table but also because the data shown will be pertinent to a subsequent discussion.

		J.D.	*Sun*	*Mercury*	*Venus*	*Mars*	*Jupiter*	*Saturn*
1958 Jan.	26	6230	306	284	311	264	210	262
1958 Feb.	5	6240	316	298	305	271	211	263
1959 Aug.	29	6810	155	138	159	175	234	271
1959 Sept.	8	6820	165	156	153	182	235	270

The entries represent the angular positions at which we on Earth see the planets in this plane on any given date. If the earth and clouds were transparent, and the sun did not blind us when we looked in its direction, we should see all the planets in a plane, very closely, and at all times in the same plane with respect to the fixed stars. They appear, at least in this discussion, as points, and we cannot tell directly how far away they are. We can specify only the angle at which we see them. Are they really moving around our earth, as they seem to be?

We must now turn to a more detailed examination of the data in Table II. The first column gives us a series of dates, starting on September 14, 1954. The dates come at ten-day intervals. Since the months have irregular lengths, the dates do not have a regular sequence. It is somewhat of a job to calculate the date of a day 280 days from now, or 3420 days ago. To simplify the counting, astronomers have adopted arbitrarily a numbering of days starting at noon, Greenwich mean time, on January 1, 4713 B.C.! The choice may seem arbitrary and peculiar, but at least it is definite. It turns out that September 14, 1954, is Julian Day (or J.D.) 2,435,000. It is now easy to find the designation of a day 280 days later. It is J.D. 2,435,280, and this turns out to be, according to our table, June 21, 1955. Now let us examine the numbers representing the angular positions of the six celestial objects listed, the sun and the planets Mercury, Venus, Mars, Jupiter, and Saturn. A few items stand out. With very few exceptions, the sun advances either 9° or 10° every ten days. We might assume that the sun moves by some definite amount every day, and that this amount lies between 9° and 10°. But then, how could we account for the fact that the table specifies 11° for the ten-day period November 27 to December 7 in 1958? Of course, we might suppose that this is an error in taking down the data. But is it? Let us assume that the observer is very sure of himself, and guarantees that, although there may be errors of a few minutes of arc in his observations, there is no error as great as thirty minutes, or $\frac{1}{2}°$, and that, therefore, the number of degrees listed is wholly reliable. If this is so, we must conclude that the motion of the sun cannot be perfectly regular. It must vary from time to time. The next question might be whether it varies regularly or irregularly.

But before going too deeply into the motion of the sun, it might be better to get an over-all impression of all the data. For this purpose a graph is indicated. One can see patterns much better in a picture than in a list of numbers. Plotting the graph is somewhat of a job, but it is very much worth doing, at first roughly in order to see the main features, and then more accurately in those areas that seem interesting.

In Fig. 17a we have plots of the motion of the sun, Mercury, and Venus, and in 17b, Mars, Jupiter, and Saturn. Perhaps few readers will appreciate what an extraordinary spectacle is revealed in these curves. Most scientists, when they try to interpret some phenomenon of nature for the first time, have only scattered observations, and those of only limited reliability. You might look at *The Watershed,* by

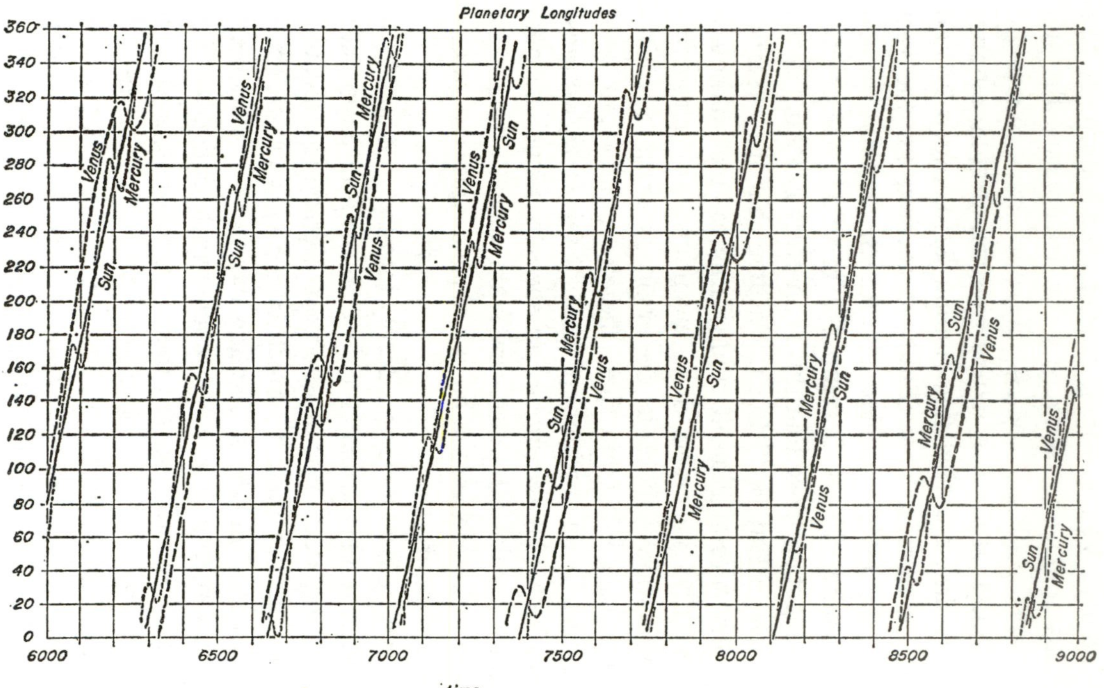
Planetary Longitudes
360
340
320
300
280
260
240
220
200
180
160
140
120
100
80
60
40
20
0
6000
6500
7000
7500
8000
8500
9000
time
Venus
Sun
Mercury

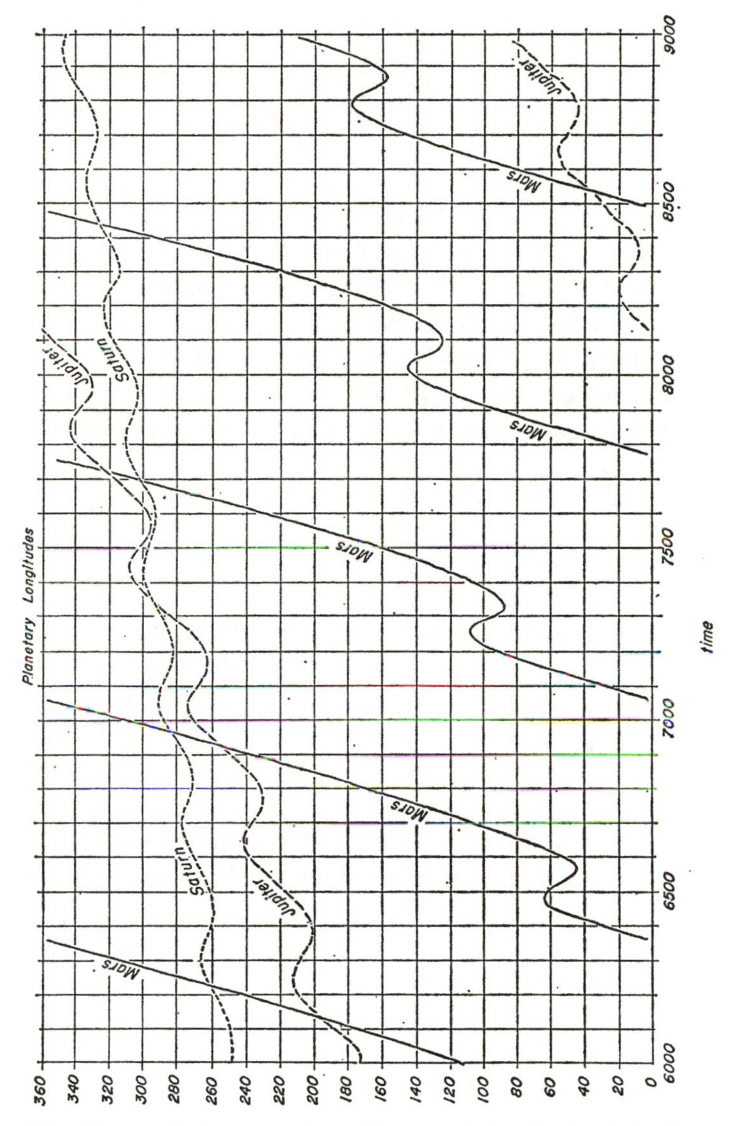

Fig. 17 (a to left, b to right). Geocentric longitude of the Sun and planets as given in Table II (Appendix).

Arthur Koestler, Volume S16 of the Science Study Series. It will give you an idea of the struggle Kepler had to get even a few reliable facts about the motion of Mars from the only man of his time with adequate instruments, the wealthy, government-supported Tycho Brahe. Kepler would have hit upon his famous laws of planetary motion much more readily with the data of Fig. 17. In fact, it is quite probable that anyone with a slight knowledge of geometry and trigonometry, and some persistence, might independently be led to some of Kepler's conclusions. You might try it. The rest of this chapter is really just a matter of providing a few slight hints, plus a little dog work.

The general features of the motion of the planets, their tortuous paths in the sky, were of course known long before Kepler and Tycho Brahe came on the scene, even long before Copernicus suggested a sun-centered solar system. They clearly led to ideas of wheels within wheels going around various axes at various rates. The more accurate observations became, the more cycles and epicycles were needed to account for them. But the one idea of the planets' moving around the sun in nearly circular orbits has taken the place of all this. As seen from the earth, the planets all seem to move to and fro. Surely the planets don't stop dead in their tracks and then back up without some tremendous forces and disturbances, as indicated by their *apparent* motion. We must now prove that their true motion around the sun in space is quite different.

The Motion of the Earth around the Sun

The great simplification in the description of the motions of the planets comes about, as we now all know, by assuming that the earth and planets move around the sun. In the celestial sphere the position of the earth as seen from the sun is diametrically opposite the position of the sun as seen from the earth. We can, therefore, plot the position of the earth with respect to the sun if we assume, for the present, a fixed separation. This has been done in Fig. 18, using the positions and dates of Table II.

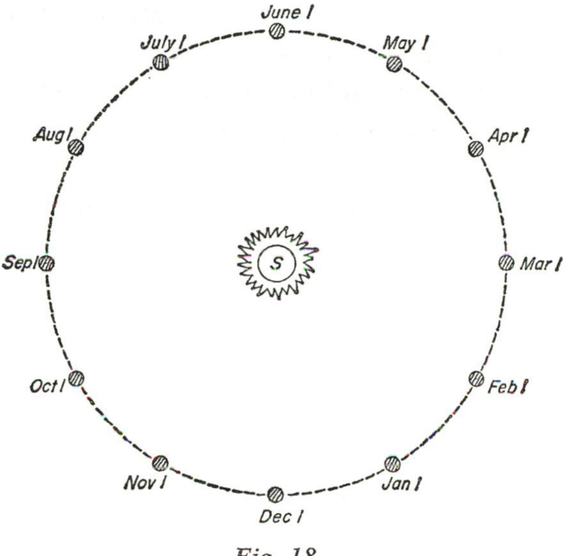

Fig. 18.

We must next examine the assumption that the earth is moving around the sun at a fixed distance. If this were so, we should expect its angular velocity, or the number of degrees traveled per unit of time, to be constant, since at a constant distance the physical conditions at every part of the orbit would be the same, and there would be no reason to expect the earth to move faster at one point of its orbit than at another. But if the distance were not constant, then the earth would be approaching, or falling toward, the sun in some parts of its orbit, and climbing away from it in others. We should expect it to gain speed as it approaches the sun and lose speed as it recedes. In other words, if the orbit were not circular, we should expect the angular velocity of the earth to be greatest when it is nearest the sun and least when it is farthest away. This hypothesis we can test, using the given data, at least to the accuracy of the data.

Since the data are not accurate enough to show a regular progression in ten-day intervals, but jump around between 9° and 11°, let us instead use a somewhat larger interval, say fifty days, and plot the observed motions during the year 1957. In the fifty days from June 10 to July 30 the sun moved from 79° to 127°, or 48°. Its average angular velocity, usually written ω (omega), is therefore 48° per fifty days. This figure has been entered halfway between these dates in the time scale of Fig. 19. The data

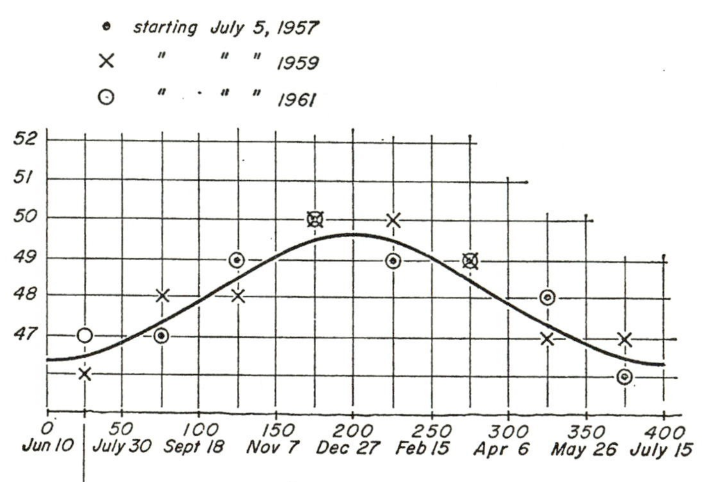

Fig. 19.

indicate an annual variation, with higher angular velocities in the winter than in the summer. To check the point, similar entries are made for 1959 and 1961. There does seem to be a slight but definite seasonal variation. From this we can estimate how nearly constant the distance from the earth to the sun is. We shall need a little math and physics, but not much.

A curve has been drawn through the points in Fig. 19 to indicate a reasonable guess as to the true variation of apparent angular velocity of the sun in the heavens, or, if we assume that the sun is fixed and the earth is rotating around it, the curve represents equally the angular velocity of the earth around the sun. It varies only a little. The ratio of the maximum to the minimum values is $49.5/46.2 = 1.07$. The variation is 7 per cent. The velocity is greatest in December, least in June.

Here are facts. Can we conclude anything more definite from them? We might consider that we have gropingly found an area that seems worth studying, but without help we are blind to the significance of these facts. We need the guidance of the principles of mechanics. These laws can act like searchlights to illuminate the interesting area we have discovered. The raw data in Table II mean little. When plotted up as in Fig. 19, they mean a little more. But if we know enough physics to apply the principle of conservation of momentum, we can find out even more.

Momentum is not an uncommon word even in non-technical language, but it may be useful to some readers to review first qualitatively, then quantitatively, just what momentum means in the science of mechanics. It is related to motion, not just the picture of motion that we see with our eye, but a more physical aspect, which we can feel when we are hit by a ball. A baseball or a tennis ball dropped from a cliff will acquire nearly the same speeds when they hit the ground at the bottom. To the eye, their motions are similar. But if you try to catch them, you can discover a great difference. There are, of course, various aspects to the difference, but the simplest and most fundamental one is what Newton called the "quantity of motion," or what we now call the momentum. It is the product of mass and velocity. In rotating systems there is a similar quantity called the angular momentum, as opposed to the linear momentum of the falling balls. The angular momentum of a mass m moving around a circle of radius r with angular velocity ω is $mr^2\omega$. The conservation of angular momentum stipulates that in an undisturbed system, or one that is not "disturbed" by forces applied from outside, this quantity remains constant, or that it must always have the same value for the earth as it goes around the sun. In particular, it must have the same value in June as in December.

If the angular momentum of the earth in moving around the sun in June is equal to its angular momentum in December, we must be able to write the following equation:

$$m_e r_e^2{}_{\text{June}}\ \omega_{e\ \text{June}} = m_e r_e^2{}_{\text{December}}\ \omega_{e\ \text{December}}$$

or, dividing both sides by $m_e r_e^2{}_{\text{December}}\ \omega_{e\ \text{June}}$, and taking the square root,

$$\frac{r_{e\ \text{June}}}{r_{e\ \text{December}}} = \left(\frac{\omega_{e\ \text{December}}}{\omega_{e\ \text{June}}}\right)^{\frac{1}{2}} = \sqrt{1.07} = 1.035$$

So, with this extra knowledge about angular momentum, and the data about the apparent motion of the sun in Table II, we conclude that the radius of the orbit of the earth varies by approximately $3\frac{1}{2}$ per cent.

Another important question to face is to determine how far it pays to go in examining data. For example, have we squeezed out all the information there is in those numbers about the apparent motion of the sun? Could we hope to find out something about other unknown quantities? What is the mass of the sun? Or of the earth? What is the average distance between the two? Would the application of another of the great principles of mechanics, the principle of conservation of energy, help us to see more deeply into the physical constitution of the earth-sun system?

Perhaps this suggestion might be stimulating to some. The potential energy of a ball of mass m held at a height h above the surface of the earth is mgh, g being the acceleration of gravity due to the earth. When the ball, starting from rest, is dropped from a height h_1 to a lower height h_2, the potential energy which it loses, $mg\ (h_1 - h_2)$, is converted into kinetic energy of motion, $\frac{1}{2}mv^2$, v being the velocity or speed of the ball when it reaches h_2. Mathematically this is expressed by

$$mg\ (h_1 - h_2) = \tfrac{1}{2}mv^2$$

If the ball starts at h_1 with a velocity v_1 and ends up with a velocity v_2, the change in kinetic energy is $\frac{1}{2}mv_2^2 - \frac{1}{2}mv_1^2$, or the final kinetic energy minus the initial kinetic energy.

$$mg\ (h_1 - h_2) = \tfrac{1}{2}m\ (v_2^2 - v_1^2)$$

Notice that the mass of the ball cancels out of this expression, but the magnitude of the acceleration of gravity, g, remains:

$$g\ (h_1 - h_2) = \tfrac{1}{2}\ (v_2^2 - v_1^2)$$

Might the same idea apply to the earth in its motion around the sun? Can we consider the earth as an experimental ball that can provide us with information about the gravitational attraction of the sun, and therefore about its mass?

A curious mind delights in examining a situation of this kind in further detail. As a matter of fact, there is an enormous amount of "gold" in Table II. We shall, however, confine ourselves to just a few points, especially regarding the purely geometrical aspects of the data. Let us leave the earth-sun system at this point and take up the planet Venus.

The Motion of Venus

If the planets are moving around in more or less circular orbits, it should be particularly simple to discover the radius of the orbits that lie between the sun and the earth's path. Consider the geometry shown in Fig. 20, representing a conceivable orbit

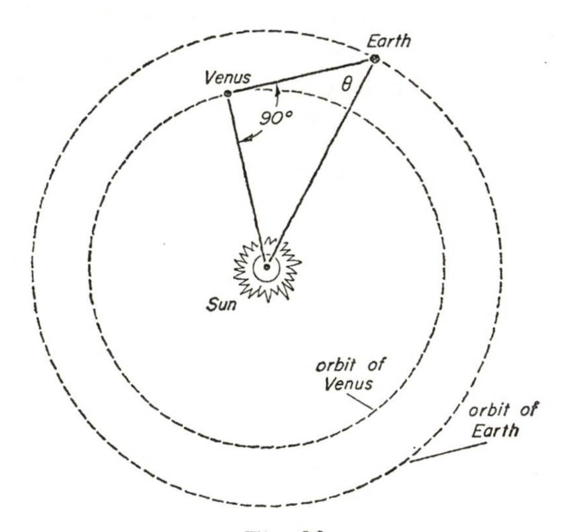

Fig. 20.

for Venus, circular and smaller than the earth's. It is clear that the maximum angular separation between the sun and Venus occurs when the line from the earth to Venus is tangent to—that is, just touches—the orbital path of Venus. In these circumstances the line from the sun to Venus is at right angles to the line earth-to-Venus. But the angle θ_{max} can be obtained from the data of Table II. In Fig. 21 the angular separation of the sun and Venus is plotted for the period June 1957 to March 1960. We see from this graph, and more accurately from Table II, that the maximum separation, while perhaps slightly variable, seems to lie between 46° and 47°. Variations of this kind would be expected if the orbit of Venus were not perfectly circular. It might be important in this connection to plot the available data for a longer period of time, to see whether greater deviations occur. It turns out that the orbit of Venus is about as nearly circular as the earth's orbit. Let us adopt an average maximum angular separation of Venus and sun as 46.5°. If we now draw the line earth-Venus at 46.5° to the line earth-sun, as in Fig. 22, we should find that, following the geometry of Fig. 20, the distance sun-Venus is about .725 times the sun-earth distance. Another way of finding this result would be to use trigonometric tables.

$$\sin \theta_{max} = \frac{r_{Venus}}{r_{earth}}$$

$$\sin 46.5^\circ = .725 = \frac{r_{Venus}}{r_{earth}}$$

An interesting question arises about Fig. 21. Why is this curve so unsymmetrical? According to this

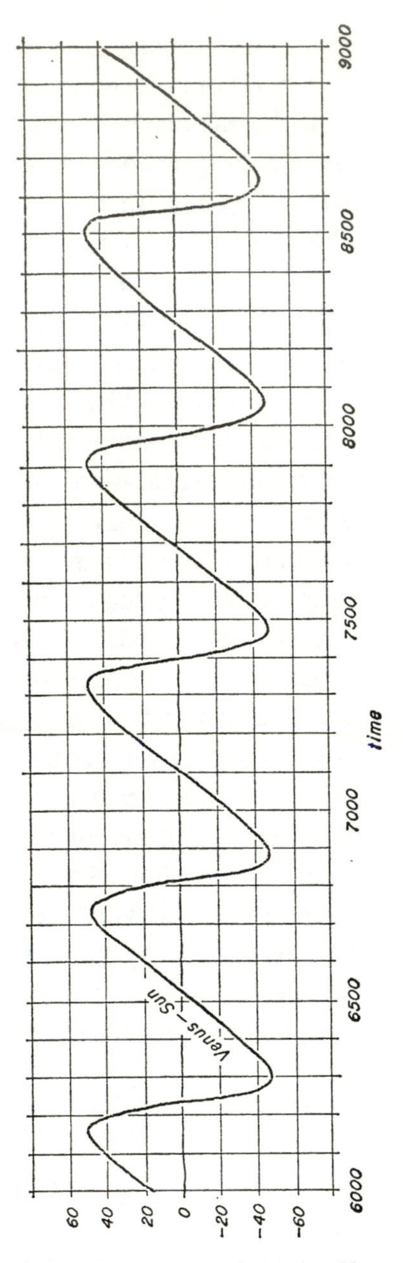

Fig. 21. The angular separation between Venus and the sun for the period June 1957 to March 1960.

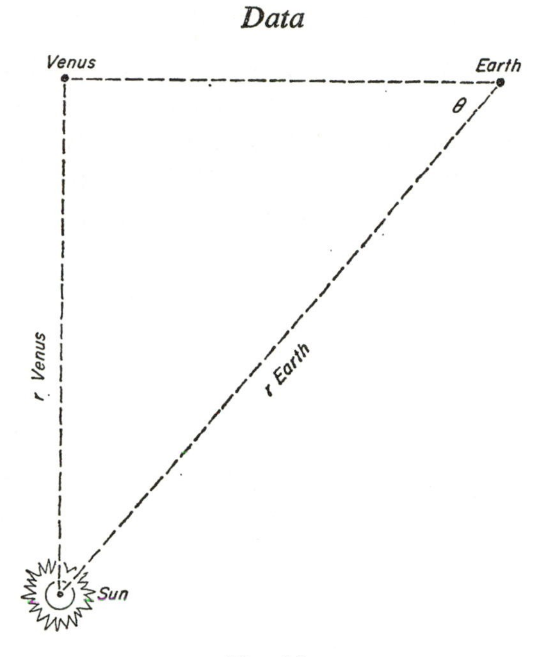

Fig. 22.

figure, Venus passes across the face of the sun alternately faster and more slowly. Why should this be? This requires some reflection, but in the end the explanation boils down to the fact that an express train roaring past your station platform at sixty miles an hour seems to be going much faster than a train seen across the valley. The greater angular speed of Venus (greater to us, that is) occurs when it is between the earth and sun; the speed is less when Venus is farther away on the other side of the sun.

Another point to consider is the length of time it takes for Venus to complete its journey around the sun. Can we calculate this from the data? In other words, can we calculate the length of a year for Venus?

From Table II we can calculate two successive

dates on which Venus is directly between† us and the sun. An example would be on about January 29, 1958 and August 31, 1959. Let us draw the corresponding situation as in Fig. 23. We know the angle ϕ between the radii describing the two coincidences.

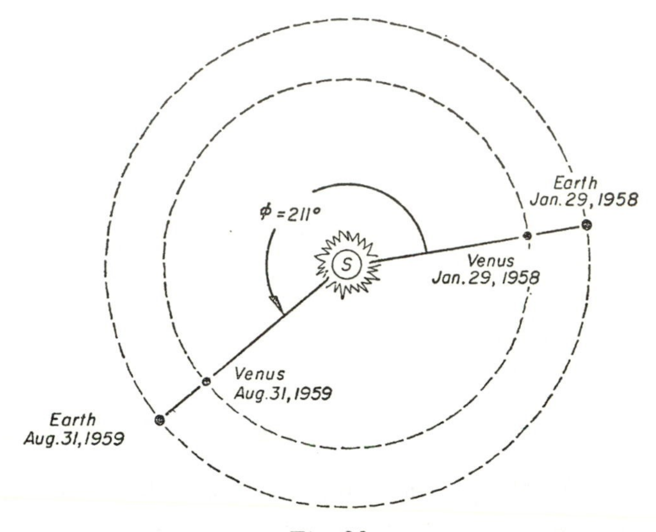

Fig. 23.

The earth moves through 360° in approximately 365 days, or at the rate of $\frac{360}{365} \cong .987°$ per day. In the 579 days between January 29, 1958, and August 31, 1959, the earth moved through 360° plus 214 × .987 or one revolution plus 211°. Therefore, referring to Fig. 23

$$\phi = 211°$$

Next we need to determine whether the earth is

† This alignment is called "inferior conjunction," as opposed to "superior conjunction," when the Earth, Venus, and the sun would be in line also, but with Venus on the far side of the sun. You will see in Table II that superior conjunction occurred on November 12, 1958, but we are concerned here with successive inferior conjunctions.

moving faster or slower than Venus in angular velocity. The total time elapsed between these two dates is 579 days. If Venus is moving more slowly than the earth, we might assume that it had moved only through the angle ϕ. In other words, on the starting date the earth moved ahead and had to go through a little more than one and one half revolutions before overtaking Venus again. At this rate the angular speed of Venus would be 211° in 579 days or

$$\frac{211}{579} = .365°/\text{day}$$

and the Venus year would be the time required to cover 360 degrees, or 990 days, since $990 \times .365 = 360$.

But it is equally possible that the angular velocity of Venus is greater than that of the earth, so that while the earth went about one and one half times around the sun, Venus went about two and a half times around. This would mean that it had traveled $360° + 360° + 211° = 931°$ in the elapsed 579 days, and that its angular speed was

$$\frac{931}{579}\ \text{days} = 1.61°/\text{day}$$

Then the length of the Venus year would be 224 days, since $224 \times 1.61 = 360°$.

Which of these answers is correct? The data are sufficient to show that Venus is moving faster than the earth. Can you convince yourself that this is so?

Another check which may be of interest comes from one of Kepler's laws of planetary motion: that for all the planets the product of the radius cubed and the angular velocity squared is constant. This is very easy to show for circular orbits. The centripetal force required to make a mass move in a circular orbit of radius r, with a linear speed v (or angular speed $\omega = \frac{v}{r}$) is centripetal force $= m\frac{v^2}{r} = mr\omega^2$. This is the force which a string of length r must exert on a stone of mass m tied at one end, and whirled around with angular velocity ω.

If this force is supplied by the gravitational attraction of the sun (mass M_s), it must be that the force of gravity

$$F_{\text{gravity}} = \overline{K}\frac{M_s m}{r^2}$$

is equal to the centripetal force. $\overline{K}$ is a constant whose magnitude depends on the units used. We must have for any planet

$$\overline{K}\frac{M_s m}{r^2} = mr\omega^2$$

$$\overline{K}M_s = r^3\omega^2$$

Since $\overline{K}M_s$ is the same for all the planets, Kepler's law, in particular for the earth and Venus, follows:

$$r_e^3\ \omega_e^2 = r_{\text{Venus}}^3\ \omega_{\text{Venus}}^2$$

or

$$\left(\frac{r_{\text{Venus}}}{r_{\text{earth}}}\right)^3 = \left(\frac{\omega_e}{\omega_{\text{Venus}}}\right)^2$$

The angular velocities vary inversely as the length of the years, Y_{Venus} or Y_{earth}, so that we may also write

$$\left(\frac{r_{\text{Venus}}}{r_{\text{earth}}}\right)^3 = \left(\frac{Y_{\text{Venus}}}{Y_{\text{earth}}}\right)^2$$

But the ratio of the radii we already have determined to be .725, and the earth's year is 365 days. We, therefore, have that the length of the year for Venus is

$$Y_{\text{Venus}} = (365)\ (.725)^{3/2}$$
$$= 365 \times .62$$
$$= 225 \text{ days}$$

which, to the degree of accuracy of our calculations, checks with the previously obtained 224 days.

The Motion of Mercury

The use of facts is generally not so simple and straightforward as the foregoing pages may indicate. The real business of digesting and assimilating data is usually much more involved. In fact, retracing the paths of pioneers is likely to be taxing and time-consuming. But in concluding this "Cook's tour," it would be a shame not to indicate a little more accurately what coming to grips with raw data means. So we shall now take this sight-seeing bus a little further from the familiar area of well-laid-out streets and signposts into the less well explored, and perhaps unfamiliar, territory.

Mercury, like Venus, follows a path in the sky that always remains close to the sun. We might attempt to treat it as we did Venus's path. The angular separation of Mercury and the sun is plotted in Fig. 24. Superficially, this resembles the corresponding plot for Venus in Fig. 21. There is a similar, generally periodic, aspect to the graph. We can recognize the more rapid passage across the sun when the planet is on the earth's side of the orbit. But the amplitude of the motion is far from constant. It was shown in Fig. 21 that for Venus the maximum deviation is always the same. This is because the orbit is almost circular and has the same angular spread when seen from any point of the earth's nearly cir-

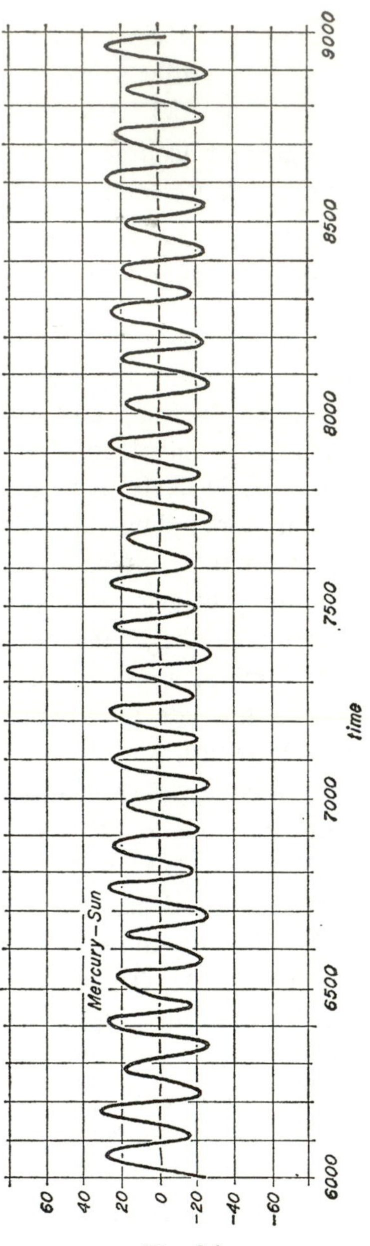

Mercury—Sun
time
6000
6500
7000
7500
8000
8500
9000
60
40
20
0
-20
-40
-60

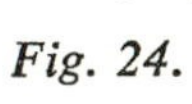
Fig. 24.

cular orbit. The orbit of Mercury, as described in Fig. 25, is clearly not circular. The angular spread as seen from the earth varies, and depends on the position in our orbit from which we view it. The problem of laying out the orbit of Mercury is consequently not so simple.

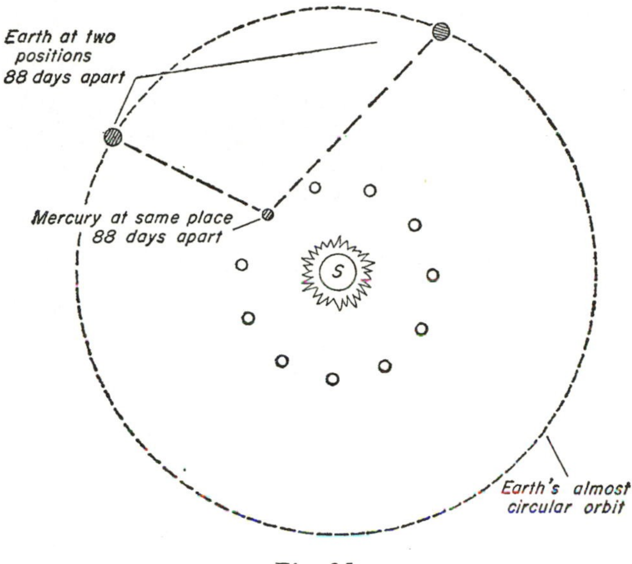

Fig. 25.

It is at this point that a little courage and perseverance are needed. If the problem of locating a distant moving object from a moving platform seems too difficult, a good step may be to simplify the problem, and to get useful ideas from methods developed in the simplification. Let us suppose that we wish to locate a protruding rock in a lake. We want to mark its location on a map—accurately. From some identifiable point on the shore in Fig. 26, we can measure, with a suitable sighting device, the angle θ_1, between a north-south line, and the line of sight to the rock.

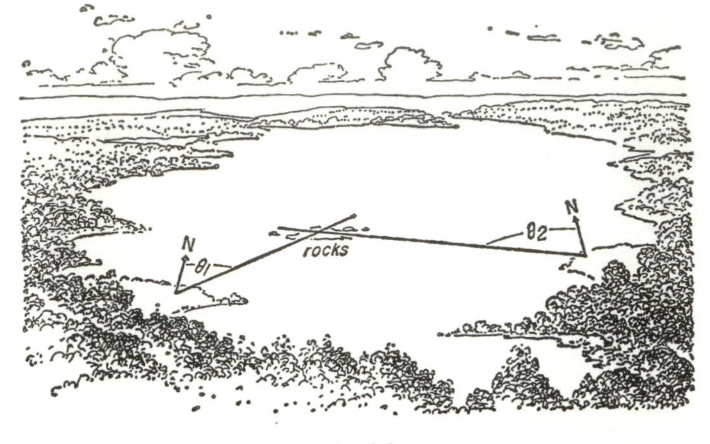

Fig. 26.

If we do not know how far away the rock is, we cannot yet pinpoint the rock. If, however, we make a second similar angular measurement θ_2 from another identifiable point, we can conclude that the rock must be at the intersection of these two lines of sight.

Now, if Mercury were not in motion, we could apply the same method by sighting on the planet from two different points, *A* and *B,* in the earth's orbit. The final solution of the problem depends, then, on getting around the difficulty of the motion of Mercury. It should not be assumed without justification that the motion of Mercury is periodic, in the sense that its orbit is repeated year after year. But if it is periodic, then our problem is solved. We have only to derive the length of the year of Mercury as we did for Venus and find out from our data where the earth was in its orbit on two dates when Mercury was at the same position. Since we already have done this once, we will forego a repetition here and merely give the result, which the reader can check for himself. The length of the year of Mercury turns out to be only 88 days.

Assuming then, subject to later verification, that the orbit is strictly periodic, we can locate it at any one position, say on January 1, 1958, by drawing lines from the earth in the recorded direction of Mercury, and similarly 88 days later. We could further locate Mercury at eight points in its orbit by choosing fixes at eleven-day intervals, say January 1, January 12, January 23, February 3, February 14, February 25, and so on. In order to locate Mercury on these dates, we must, of course, note its position as seen from the earth, not only on these dates, but one Mercurian year, or 88 days later.

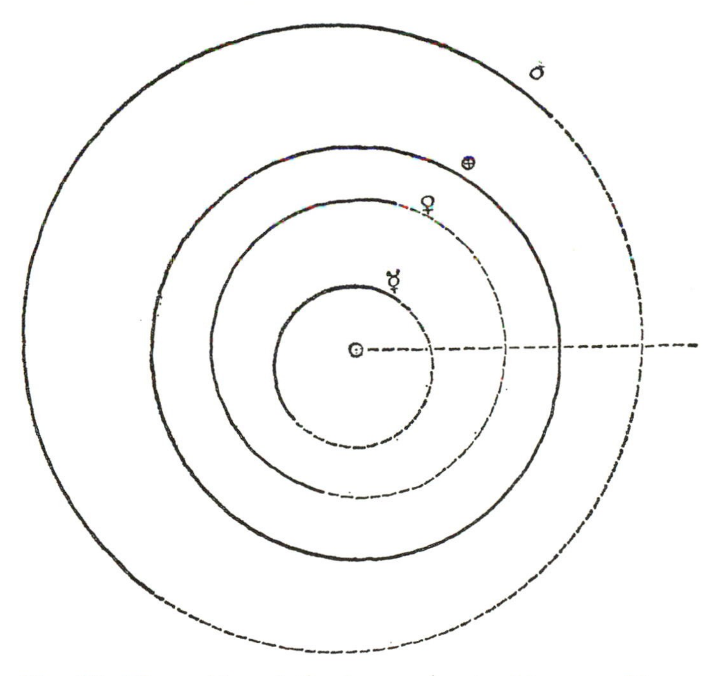

Fig. 27. The orbits of the inner planets: Mercury, Venus, Earth, and Mars. Planets are designated by ancient symbols derived from Latin and Greek mythologies.

Such a procedure would lead to the orbit of Mercury around the sun plotted in Fig. 25. At first glance this may seem to be a circular figure, with the sun off center, but actually it is an ellipse, with the sun at one focus. What is an ellipse? If you are interested in geometry you doubtless know. If you are interested in astronomy you will doubtless find out. The orbits of all the planets are ellipses (Fig. 27) with the sun at one focus. These are the orbits that follow from Newton's law of gravity. They can be computed from the data in Table II. As the foci of an ellipse approach each other, the ellipse approaches the form of a circle, and in this sense the planets with nearly circular orbits conform to the previous statement.

CHAPTER 3

Analysis

MATHEMATICS—THE LANGUAGE OF PHYSICS

Have you ever stopped to think how important language is? Language is a means of communication, of course. But it is far more than that. It is an indispensable ingredient of thought itself. You could not plan your day without using unspoken words to help to abbreviate and summarize—cornflakes, run, homework, telephone, flat tire, money, study, grandmother, haircut . . . And the kinds of thoughts a person can have depend to an enormous extent on the words he knows and his skill in using them. Clearly, a person who knows only simple words, like those strung out above, cannot enter into those parts of life dealing with voltage, current, impedance, reactance, inductance, and capacity, or resonance, F sharp, G flat minor, fugue, treble clef, percussion, allegro, vivace, strings . . .

Language, in its broadest sense, makes use of symbols other than words. For example,

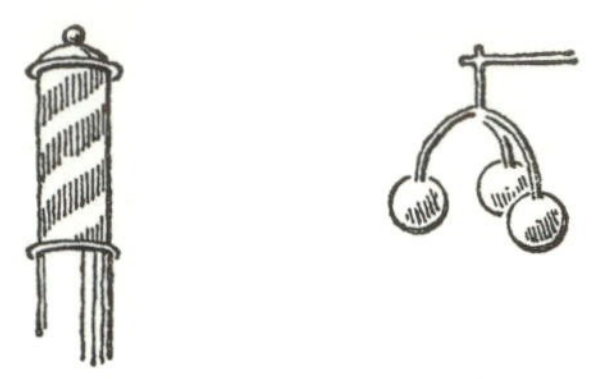

are such symbols, referring to professions, or

are symbols referring to nationalities, or political regimes. In quite a different category are the symbols used in describing electrical circuits:

These symbols are an indispensable part of the language of electrical engineering. Without a wiring diagram it would be impossible to describe even the simple set of connections that brings electricity to the switches and outlets of your house. Words would be most clumsy.

At first glance this must seem strange. How can the ordinary means of communication and expression be superseded? How could a report on the awful consequences of war be better expressed than in this simple language from the first page of John Hersey's *Hiroshima?* (See Fig. 28.)

I · *A Noiseless Flash*

AT EXACTLY fifteen minutes past eight in the morning, on August 6, 1945, Japanese time, at the moment when the atomic bomb flashed above Hiroshima, Miss Toshiko Sasaki, a clerk in the personnel department of the East Asia Tin Works, had just sat down at her place in the plant office and was turning her head to speak to the girl at the next desk. At that same moment, Dr. Masakazu Fujii was settling down cross-legged to read the Osaka *Asahi* on the porch of his private hospital, overhanging one of the seven deltaic rivers which divide Hiroshima; Mrs. Hatsuyo Nakamura, a tailor's widow, stood by the window of her kitchen, watching a neighbor tearing down his house because it lay in the path of an air-raid-defense fire lane; Father Wilhelm Kleinsorge, a German priest of the Society of Jesus, reclined in his underwear on a cot on the top floor of his order's three-story mission

Fig. 28.

But look at Newton's second law of motion, first in English,

> "Rate of change of quantity of motion is proportional to the applied force, and takes place in the direction of the force. . . . Quantity of motion is proportional to mass and velocity conjointly."

Then see it expressed symbolically, in mathematical form, $F = ma$ or, spelled out in another more detailed way,

$$\vec{F} = \frac{d}{dt}(m\vec{v})$$

Most readers will not be familiar with the notation, or with the ideas behind it, but they will at least recognize a certain elegance in its brevity.

There are, in fact, many means of expression, and they are not at all equivalent. Take, for example, Albrecht Dürer's drawing of his mother (Plate III). Surely it would be impossible to say just this in words. Or again, take the two pages of the opening of the second movement of Beethoven's Ninth Symphony (Plate IV). If you are not familiar with it, you should listen to it, and follow on the printed page the unfurling of its meaning. This could hardly be described in words. Nor could this:

$$\vec{\nabla} \times \vec{E} = -\dot{\vec{B}}$$

$$\vec{\nabla} \times \vec{H} = \dot{\vec{D}}$$

$$\vec{\nabla} \cdot \vec{E} = 0$$

$$\vec{\nabla} \cdot \vec{B} = 0$$

which is James Maxwell's famous description of the electromagnetic properties of a vacuum! These symbols contain essential knowledge about how the earth is warmed by the sun, how we can find our way across the street, and how sounds and pictures can be transmitted to us from distant places.

The essentials of the abstract general truth in a physical situation can be described in mathematical terms incomparably better than in other ways. This description is unlike an exact photographic reproduction of reality. It is in no sense a record or a moving picture. A physical law mathematically described reveals certain general essentials, not just any one particular sequence of events. But it can be used to describe and predict actual physical events.

Mechanical Vibrations

In order to show this we propose to study simple motions, particularly oscillations. Perhaps, once we have described some of the ideas that go into an understanding of oscillations, once we have begun to see what is common to all oscillations and what is particular to this or that or the other oscillating system, we shall have laid a groundwork of thoughts on which to base a mathematical generalization. It is, of course, clear that our treatment must be relatively superficial. A good many years are needed to make headway in mastering even simple aspects of mathematical physics, just as many years are needed to give a student of music an insight into a musical score. Nevertheless, it can be a very thrilling experience for a person unfamiliar with music to have a musician take him to the piano and indicate, in however sketchy a fashion, some of the high points in an orchestral or operatic work–to reveal in a simplified way the beauty of the musical ideas involved. Similarly, it may be that some of you will be attracted to the kind of abstraction concerning vibrational motion contained in a mathematical statement, and may go at the study of physics or your reading in the literature of science more eagerly because of this experience.

As we proceed with our abstract analysis, remember the realities involved in actual oscillations, revealed to us in quite different ways. Perhaps when you think of this word–oscillation–you think of some monotonous buzzing, the rumbling of trucks on a cobbled street when you are trying to go to sleep, or the humming sound of an electric fan. These sounds may seem dull and uninteresting, but they sometimes can be of great significance. The regular motion of the pendulum in a clock swinging to and fro measures the passage of time accurately. It is not as simple as it looks–and it certainly involves many different kinds of ideas. For example, how, really, can we measure the passage of time reliably? If we build two clocks with the greatest care and precision possible, and find that after a while they are not running in synchronism, which shall we believe? Shall we rely on the movement of astronomic bodies, and use the movement of the earth as an absolute measure? Or can we rely on clocks whose rate of speed is determined by atomic vibrations? Are these two rates accurately the same? Are astronomic clocks and atomic clocks identical in their measurement of the passage of time? These are questions about which scientists are not quite satisfied. To a certain degree of accuracy all these clocks seem to yield the same measurement of time; but if we could make them run a hundred, a thousand, a million times more reliably, would there be discrepancies which would concern the various manifestations of time itself?

Our ears make us aware of sounds–vibrations in

the air—without our being able to follow the to-and-fro motion, as in a pendulum clock. But how, then, do we recognize the quality of musical sounds? A pure tone is rather dull. How can this be changed into the more interesting sound of a note on the violin, or on the oboe? A musical note, we know, can be achieved by combining many pure tones of different frequency and different amplitude. Up to the present, the production of the beautiful sounds, of orchestral music, of anthems, of requiems, has been entirely, or almost entirely, a matter of artistry, of human experience. A violin does not have its shape because some engineer decided that this form would make possible the particular qualities of tone that actually are produced by the violin. It has this particular shape because people, through generations, have found that it produces results that musicians so very much desire—and, incidentally, because the shape allows the player to draw his bow across the strings! No one understands in detail how this particular violin tone has been achieved. However, as science and engineering progress, the production and control of sound are being considered in new ways. New ideas and new materials are leading to acoustical structures which enhance, or eliminate, reverberation. We seem to be reaching an era when people who have skill as scientists, engineers, and artists will attempt acoustical engineering in a more fundamental way than they have in the past. They are even now using knowledge of vibrations to produce sounds that no musical instrument has so far produced, and to combine these sounds to produce new kinds of music.

Some objects, like the pendulums of clocks, are rigid, and can oscillate in only one way. Others, like a violin string, can vibrate in several ways. Because the ends of violin strings are held fixed, there must always be nodes, or motionless points on the vibrating strings. The fundamental vibration is that in which the only two nodes are the end points. Harmonics are modes of motion in which there are additional nodes, equally spaced along the length of the string. Two-dimensional objects like the stretched skin on the end of a drum can be made to vibrate in more complicated ways that are difficult to see. A metal plate can be made to vibrate in many different and characteristic ways, depending on its shape, or on how it is held. The patterns of vibration can be made visible if you put sand on the plate. Wherever the plate is actually moving up and down the sand particles are jostled around until they reach the nodes, or quiet places, where there is no jostling. From Fig. 29 you can get an idea of the shapes of

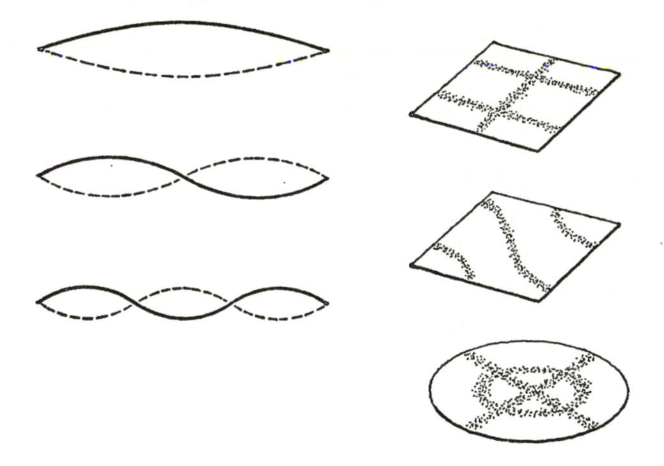

Fig. 29.

oscillations of a string with fixed ends, and of square and round plates.

Three-dimensional objects also have characteristic

modes of vibrations. Atoms are little clouds of electrons held by a positively charged nucleus, and the different states of atoms are characterized by different vibrational patterns very much as the flat plates in the illustration would be.

There is, however, a very important distinction between the vibrations of musical instruments and the vibrations of atoms. The mechanical vibrations of strings or plates push the adjacent air to and fro as they vibrate, and as a consequence sound waves are sent out. The pulsating electrons in an atom do not produce sound waves. Because of their electrical nature, however, they can emit electromagnetic waves. The range of frequencies with which atomic electrical systems can oscillate—and consequently the range of wave lengths which they can emit—is enormous. In terms of frequencies they can be as low as a few repetitions a second. We can follow these oscillations up through the spectrum to higher and higher frequencies, going first to radio waves, which oscillate millions of times per second, and then to radar and microwaves, which oscillate thousands of millions of times per second, and on through heat waves to visible light, which oscillates to and fro 10^{15} (one followed by fifteen zeros) times per second, and on through X-rays and gamma rays to cosmic rays, whose upper limit we do not yet know. And all this almost inconceivable activity that is present in space and in the air around us and in our bodies all the time, all these radiations are plowing to and fro through space in all directions and are continually being attenuated and modified by matter. They are an essential part of the activity of the universe. There is just one little band, barely an octave, which we call light, that makes it possible for us to see—to walk around, to recognize each other, to read, to study, and to observe the changing colors of the seasons. At one end of the visible electromagnetic spectrum is red light. At the other end is the violet. Everything above or below this little band is invisible to us. We cannot see it.

It is a very interesting and mysterious fact that our sense organs for detecting the mechanical vibrations in the air near us in the form of sound, and the electromagnetic vibrations entering our eye in the form of light, are quite different in a very fundamental way. If we play two notes or more on the piano simultaneously, a trained ear can detect what these notes are, and any ear at all can tell that two notes are being played and not just one. The ear is analytical in this sense. It can sort out the different frequencies of vibrations which may occur simultaneously. The eye, however, does not have this analytical quality. We can produce a green sensation not only by using light sorted out by a spectroscope and having the frequency or wave length characteristic of green, but also, for example, by mixing the yellow and blue frequencies. The eye is not analytical. It cannot distinguish pure electromagnetic waves from mixtures of electromagnetic waves.

This, in a sense, is what we shall be discussing. This is one aspect of the vibrations and oscillations without which life and consciousness would be impossible. An artist can make a caricature of a man which reveals deep-seated qualities of his character without putting in all the photographic detail of his appearance. In a similar way, with a stroke of the pen, a physicist can write a few symbols which abstract from all this variety of vibrations and waves certain elements common to them all. It is this artistry that we shall begin to explore with the help of mathematics–especially geometry, the mathematics of shape.

The Description of Oscillations

In order to come nearer to an understanding of oscillations, one must somehow get beyond the obvious facts, the mere wiggling to and fro of the tuning fork, or up and down of the fish scales. In these examples the taking of extensive and precise data gets us nowhere. It is true that a certain amount of factual information is in order. If you change the weight on a spring balance, you alter the time required for an oscillation, but if you alter the weight at the end of a pendulum, you do not. Such facts we may consider to have been observed, and these must be accounted for before we can feel that we are on the right track, that we are making progress. The real problem before us, however, is not this. It is to make use of Newton's laws of motion. How can we start to apply $F = ma$ to a pendulum?

This is a very simple task indeed if you are familiar with the differential calculus. It is almost impossible if you are not. However, we are going to try a short cut. We are going to try to present the sequence of ideas that lead to the answer to this problem. They are quite simple, and can be described geometrically, using graphs that are easy to understand. This process in a more usual formal presentation takes much longer because there are many important generalizations of these steps that must be examined carefully if you are laying the foundations of certain universally applicable ideas. It takes a long time to teach someone to read an orchestral score well enough so that he can "hear" the music, but it is quite simple to teach him to whistle a theme.

What is the theme we want to whistle? It begins with a graph, a picture of an oscillation; for instance, of vibrational motion fairly easy to follow with the eye—the motion of a weight hanging on a spring and moving up and down, let us say a few times a second. For purposes of making a record we might consider not the motion of the weight itself, but of the shadow of the weight on the wall. A useful record of the motion might be obtained if we considered the motion of the shadow on a piece of moving photographic paper. If the paper moves horizontally while the shadow spot is bobbing up and down vertically, the resultant curve would be a wavy line like that shown in Fig. 30. We can imagine making a record of this kind for any vibrating object. If our imaginary

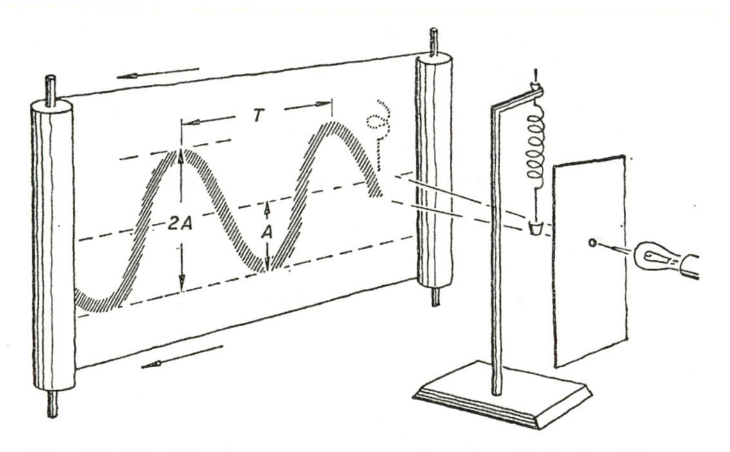

Fig. 30. The linear measurement shown as T in the diagram actually is the wavelength; it becomes the period only when it is translated into the units of time represented on the paper which is moving behind (so to speak) the shadow of the oscillating spring.

paper moves horizontally with a uniform speed, we could say that the distance along the strip is a measure of time, and the vertical position of the shadow a measure of the position of the weight hanging on the string. We would expect that records of this kind obtained from different vibrating systems would all be different. Nevertheless, there would be certain similarities, and we might attempt now to describe these similarities.

The basic characteristic of vibrational motion is that it is repetitive. Something happens over and over and over again. Even if the motion is not exactly repeated, there is a repetitive cycle. If this cycle is not present, then we are not dealing with vibrational motion. Certain words have been commonly adopted to describe this repetitive aspect of vibrational motion. The first is the *period* of the motion. This is simply the time required for the motion to repeat itself, and is designated by capital T in the figure. A related quantity is the *frequency* of the vibrational motion. This refers to the number of repetitions or cycles that occur in one second. The frequency, designated by a small f, is the inverse of the period. Thus, if the period is one second, the frequency is also one cycle per second. We would have one cycle completed in every second. If the period is two seconds, then we would have a frequency of $\frac{1}{2}$ cycle per second, and so on.

So much for the aspects of the oscillatory motion on the time scale. In the vertical scale, which measures displacement of the particle, the new word to be learned is the *amplitude* of the motion, designated by A in the figure. This quantity is generally defined with respect to a central equilibrium position. The distance between the highest and lowest displacements of the vibrating object is therefore twice the amplitude of the vibrational motion.

Another important geometrical concept in graphs of this kind is that of a relationship between the displacement and the time. The vertical displacement x is thought of as being a *function* of, or in other words depending on, the time t. If the time is given, then the graph, which is a pictorial description of this dependence or functional relation, specifies the displacement at this particular time. Notice that inherent in the information supplied in such a graph is not only the actual position at any given time but also the changing position as time elapses. The graph specifies both the position and the velocity as a function of time. To appreciate this more fully, consider how we would describe the "motion" of a point at rest on an x,t graph. This is done in Fig. 31a. We choose some arbitrary initial time for the origin O, and plot later times to the right along the horizontal axis. If the position of an object does not change with respect to time, it must be at rest. A particle at rest is therefore described in an x,t diagram of this kind as a horizontal line, for example, from A to B, and at any time the point is at the same "x." On the other hand, a particle moving with a uniform velocity is described by a straight line having some inclination to the horizontal axis as, for instance, the line OA in Fig. 31b. If the particle moves a distance of S feet in one second, and then in a second second moves the same distance S again, the point describing it will move along the straight line shown. The speed is S feet per second. If we complete a right triangle having a horizontal and a vertical leg, as for example the triangle abc, then the slope of the representative line is given by the ratio of the length ab

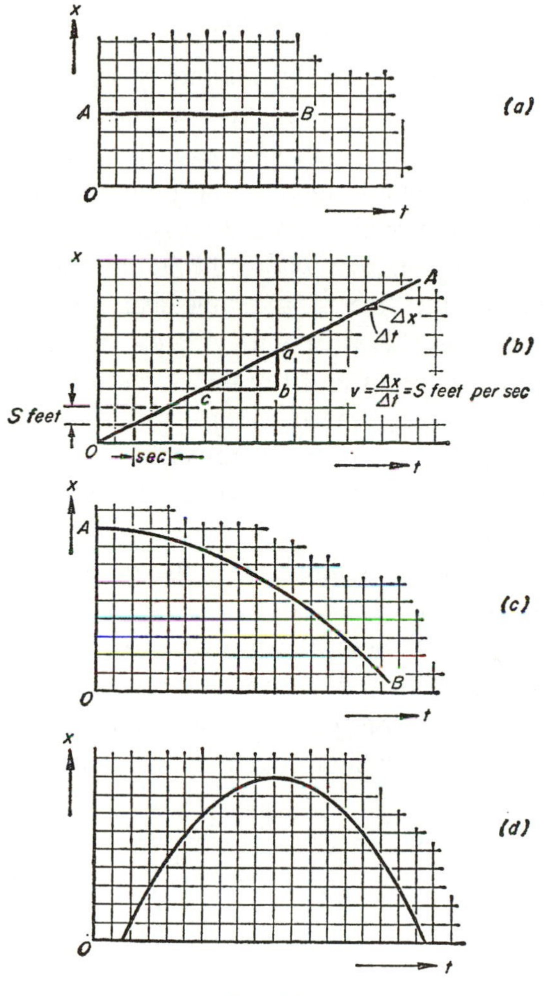
x
A
B
O
t
(a)
x
A
Δx
Δt
a
b
c
S feet
O
sec
t
$v = \frac{\Delta x}{\Delta t}$ = S feet per sec
(b)
x
A
B
O
t
(c)
x
O
t
(d)

Fig. 31.

to the length *bc*. It is the amount of rise in a given run. If the position x changes by an amount Δx in a time Δt, the speed v of the particle described is $\Delta x/\Delta t$.

$$v = \Delta x/\Delta t$$

The Greek letter delta is commonly used to designate a small change. Thus Δx is a small change in x occurring during a small change in time Δt.

A particle initially at rest and gradually falling in a downward direction would be described by the curve *AB* in Fig. 31c where the initial slope, or inclination of the line to the time axis, is zero. The slope becomes negative and greater and greater as the particle gains speed from the acceleration of gravity. And lastly, a particle thrown up in the air, straight up, gradually loses its vertical velocity until it comes to rest, and then starts to reverse its motion and fall down; its motion would be described as in Fig. 31d. The inclination, or slope of the line is at first positive, gradually approaches zero, and then becomes negative, describing downward motion. In a similarly constructed graph of vibrational motion, Fig. 32, we can interpret the curve not only in terms of positional coordinates as a function of time, but

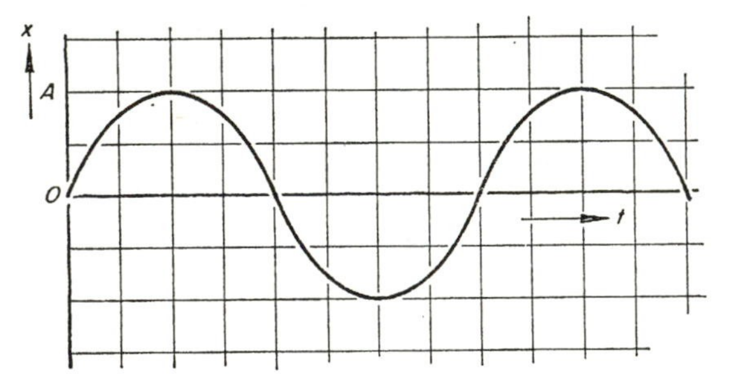

Fig. 32.

also velocity, or speed as a function of time. The fact that the curve has zero slope, or is horizontal at its vertical extremities, means that at these points the particle has come to rest. In those parts of the curve that have a positive slope in which the displacement is going from negative through zero to positive values, it has an upward or positive velocity, and in the parts of the curve that have a negative slope, that is, are sloping downhill as we move from left to right across the graph, there is a negative or downward velocity and the particle is falling.

The value of a graphic representation of motion must now be very clear. In Fig. 33 we have examples of a wide variety of oscillatory motions. These have very different frequencies and very different amplitudes. The periods vary all the way from almost twelve hours for the tides and amplitudes of twenty feet or so, through clock pendulums with periods of around a second and amplitudes of a few inches, on to the supersonic vibrations of a bat's vocal chords with frequencies probably on up to megacycles—that is, with periods of a millionth of a second and amplitudes that are more conveniently described as small pressure changes rather than linear displacements of the air.

For purposes of comparing these with each other we may change the displacement scale and the time scale so that all have the same period and all have the same amplitude. But even on this scale they all do not have the same shape. We may then ask ourselves whether there is any one particular shape that has a significance beyond all others for our purposes. It turns out that, both mathematically and physically, there is. One might think that it would be simplest to study vibrations that can be described

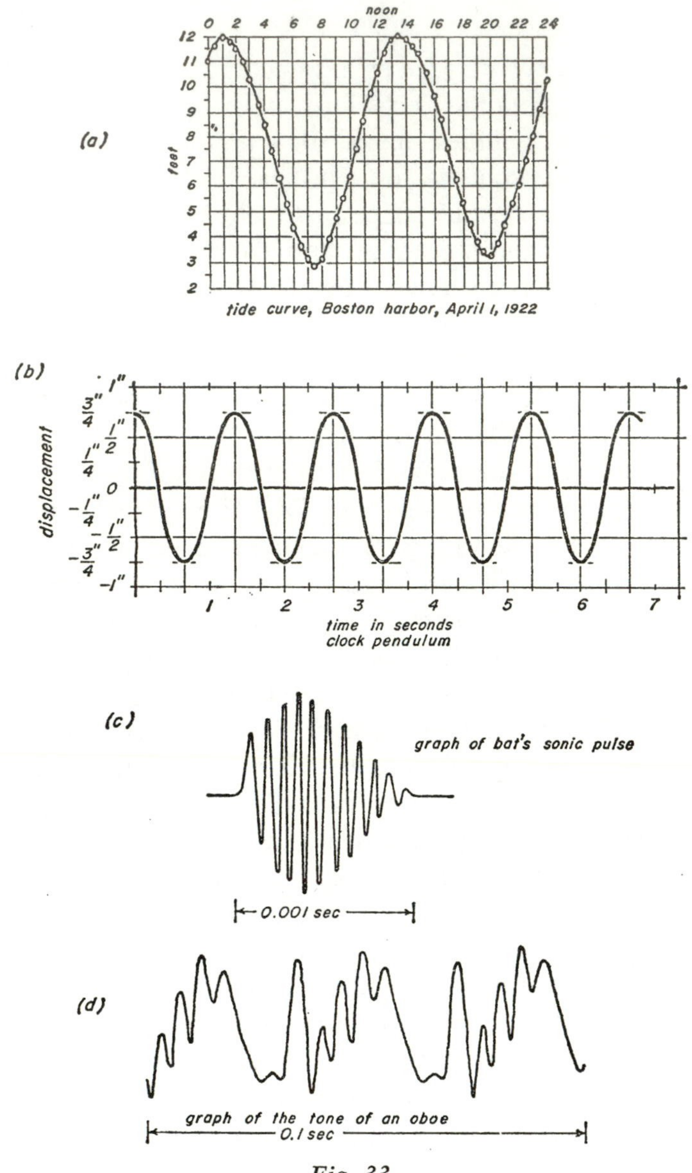

(a)
noon
0 2 4 6 8 10 12 14 16 18 20 22 24
12
11
10
9
8
7
6
5
4
3
2
feet
tide curve, Boston harbor, April 1, 1922
(b)
1"
3"/4
1"/2
1"/4
0
−1"/4
−1"/2
−3"/4
−1"
displacement
1 2 3 4 5 6 7
time in seconds
clock pendulum
(c)
graph of bat's sonic pulse
0.001 sec
(d)
graph of the tone of an oboe
0.1 sec

Fig. 33.

in terms of straight lines, like the rectangular "curves" or triangular "curves" shown in Fig. 34. But sharp corners seldom exist in nature and are hard to describe mathematically. In the vibrations with which we are concerned there are continuous changes from motion, to rest, and then motion in the other direction, rather than sudden jerky changes.

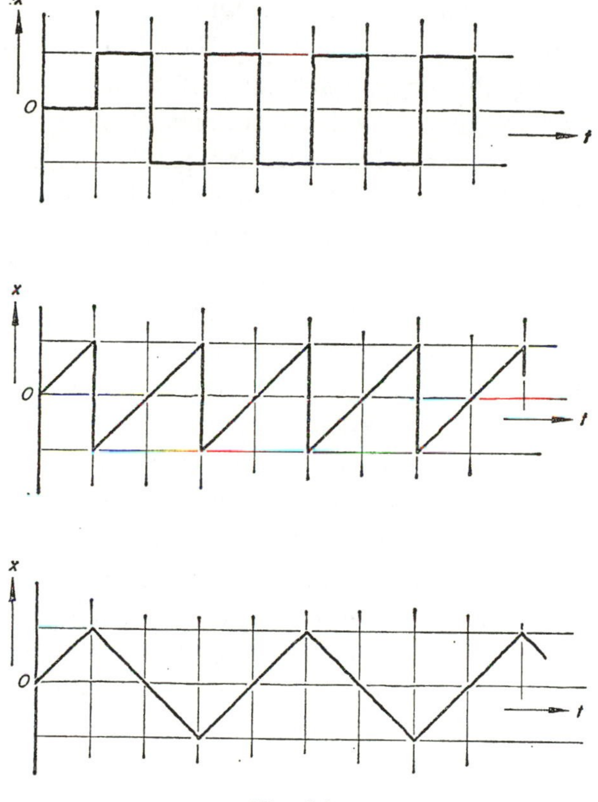

Fig. 34.

Simple Harmonic Motion

The vibrational curve form which is best suited for our study is one that is constructed geometrically from the motion of a point on the edge of a rotating circle. This is called simple harmonic motion. It can be described in terms that do not depend on any one experiment and that are universally repeatable and applicable. It can be described with arbitrary exactness, the degree actually achieved depending only on the time we take to describe it.

The reason for adopting this pure, simple harmonic motion for the description of vibrations does not depend, however, solely on the fact that we have a straightforward geometrical rule for specifying what it is, and that we avoid sharp corners. The reason goes far deeper. As we shall see further on, many actual vibrations approach this particular one very closely. The explanation underneath this fact is one of the main objectives of this discussion. There is something in physics, there is something in the objects around us, in physical laws, in the nature of forces, in the nature of mass, and in the nature of motion, that takes us to this motion of a point on the circumference of a circle rotating with uniform velocity as the key for describing vibrations. We might, for example, mount a cardboard disc so that it is free to rotate about its axis, as shown in Fig. 35. This disc is to have a knob on its edge, and it is to be illuminated with a horizontal beam of light perpendicular to the axis of the wheel. The shadow of this knob on a nearby wall, as the wheel turns around, will move up and down. This shadow will perform

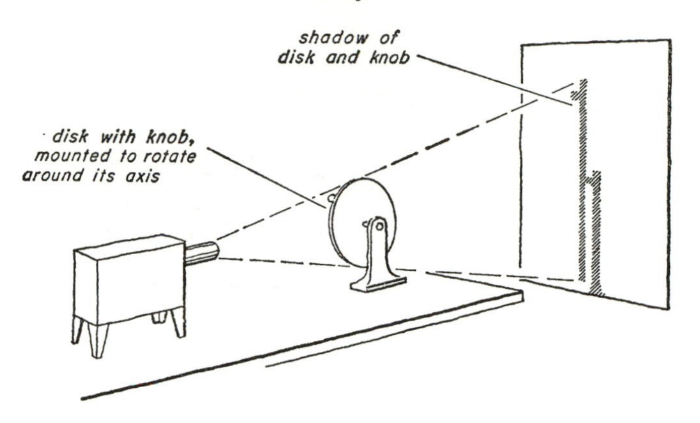

Fig. 35.

simple harmonic motion of the type we are concerned with. The motion is accurately plotted in Fig. 36.

We have now reduced the plotting of simple harmonic motion into something purely geometric in which actual motion may be eliminated altogether. We can define the vertical position of the knob on the circumference of the disc in Fig. 37 in terms of

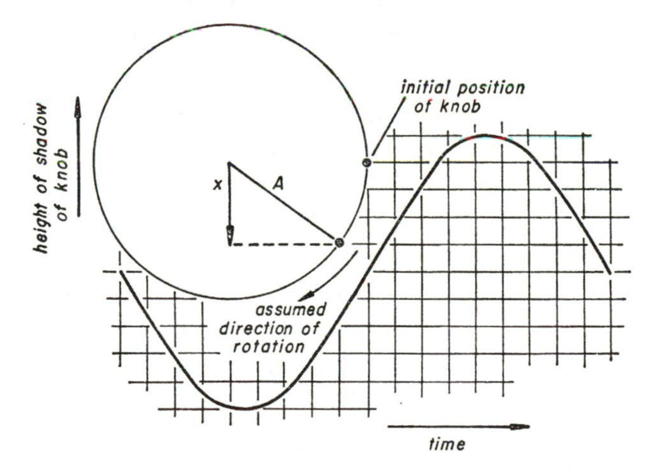

Fig. 36.

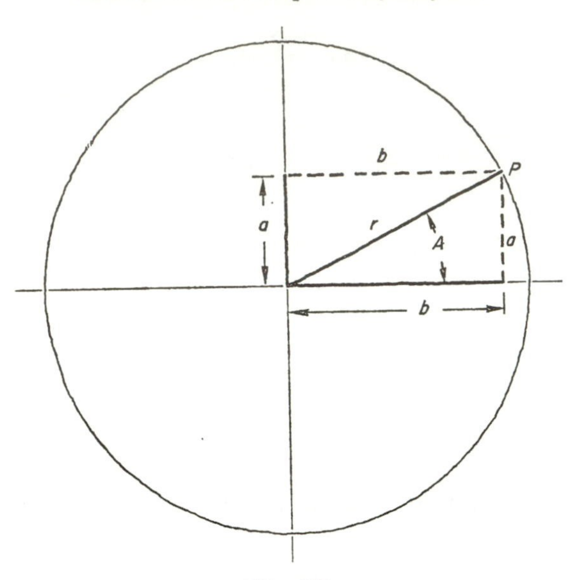

Fig. 37.

the orientation of the disc instead of the particular time at which a rotating disc has this orientation. The curve representing simple harmonic motion may then be described in terms of the vertical displacement of a point on the circumference of the disc or circle as a function of its angular position.

The Descriptive Elements of Simple Harmonic Shapes

The simple discussion we have given describing the oscillatory motion of a point on the edge of a rotating circle has in it the most diverse and fascinating ramifications. Human beings in all parts of the world have long considered these matters. For centuries mathematicians have concerned themselves with possible descriptions of these simple harmonic shapes. Astronomers and physicists have needed detailed knowledge about them for their descriptions of the material universe and for understanding of the underlying laws. Before plunging on to our main theme, it seems well to pause a moment to consider the history and implications of some of the ideas which we have outlined. They are enormously important to us all, even if they may seem trivial and academic.

In the first place there is again the question of names. In Fig. 37 we have a circle of radius *r*. The curve we have drawn in Fig. 38 is either a vertical

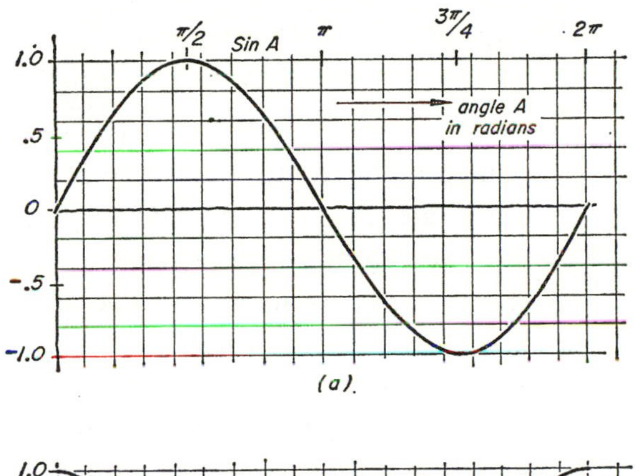

(a)

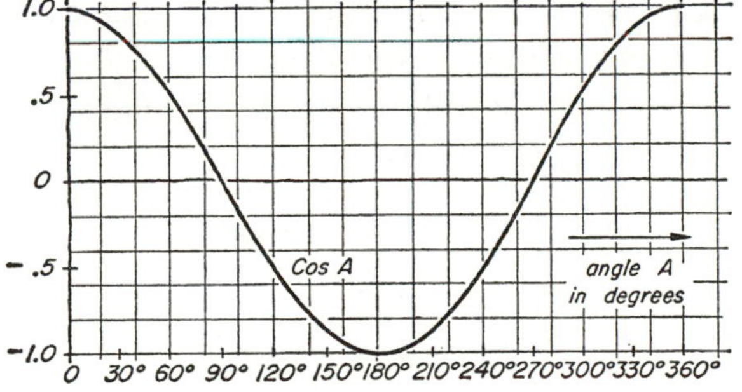

(b)

Fig. 38.

displacement of the point p on the circumference, or a horizontal displacement, depending on where we choose the starting point. The lengths of these projections are marked a and b in the figure. If we plot the lengths a and b as the angle A grows from 0 to 90 to 180 to 270 to 360 degrees, we get curves like those in Figs. 38a and 38b. In a circle of radius r, a/r is called the *sine* of the angle A and b/r is called the *cosine* of the angle A. The words sine and cosine are conventionally abbreviated to sin and cos, so that we have the following notation:

$$\sin A = \frac{a}{r}$$

and

$$\cos A = \frac{b}{r}$$

These trigonometric functions have very interesting properties which are listed in mathematical tables. A table of these functions is given below.

The properties of the sine and cosine functions can be derived from theorems in geometry. For example, $\sin^2 A + \cos^2 A = 1$. This is really only an expression of the Pythagorean theorem, which states that in a right triangle the sum of the squares of the legs is equal to the square of the hypotenuse. If we write for $\sin A$, a/r and for $\cos A$, b/r, and substitute in the above equation, we have

$$\frac{a^2}{r^2} + \frac{b^2}{r^2} = 1$$

or more simply

$$a^2 + b^2 = r^2$$

which is the theorem of Pythagoras applied to the right triangle in Fig. 37.

NATURAL TRIGONOMETRIC FUNCTIONS

Angle		Sine	Cosine	Tangent	Angle		Sine	Cosine	Tangent	Angle		Sine	Cosine	Tangent
Degrees	Radians				Degrees	Radians				Degrees	Radians			
0....	0.000	0.000	1.000	0.000	31....	0.541	0.515	0.857	0.601	61...	1.065	0.875	0.485	1.804
1....	0.017	0.017	1.000	0.018	32....	0.556	0.530	0.848	0.625	62...	1.082	0.883	0.470	1.881
2....	0.035	0.035	0.999	0.035	33....	0.576	0.545	0.839	0.649	63...	1.100	0.891	0.454	1.963
3....	0.052	0.052	0.999	0.052	34....	0.593	0.559	0.829	0.675	64...	1.117	0.899	0.438	2.050
4....	0.070	0.070	0.998	0.070	35....	0.611	0.574	0.819	0.700	65...	1.135	0.906	0.423	2.145
5....	0.087	0.087	0.996	0.088										
6....	0.105	0.105	0.995	0.105	36....	0.628	0.588	0.809	0.727	66...	1.152	0.914	0.407	2.246
7....	0.122	0.122	0.993	0.123	37....	0.646	0.602	0.799	0.754	67...	1.169	0.921	0.391	2.356
8....	0.140	0.139	0.990	0.141	38....	0.663	0.616	0.788	0.781	68...	1.187	0.927	0.375	2.475
9....	0.157	0.156	0.988	0.158	39....	0.681	0.629	0.777	0.810	69...	1.204	0.934	0.358	2.605
10....	0.175	0.174	0.985	0.176	40....	0.698	0.643	0.766	0.839	70...	1.222	0.940	0.342	2.747
11....	0.192	0.191	0.982	0.194	41....	0.716	0.656	0.755	0.869	71...	1.239	0.946	0.326	2.904
12....	0.209	0.208	0.978	0.213	42....	0.733	0.669	0.743	0.900	72...	1.257	0.951	0.309	3.078
13....	0.227	0.225	0.974	0.231	43....	0.751	0.682	0.731	0.933	73...	1.274	0.956	0.292	3.271
14....	0.244	0.242	0.970	0.249	44....	0.768	0.695	0.719	0.966	74...	1.292	0.961	0.276	3.487
15....	0.262	0.259	0.966	0.268	45....	0.785	0.707	0.707	1.000	75...	1.309	0.966	0.259	3.732
16....	0.279	0.276	0.961	0.287	46....	0.803	0.719	0.695	1.036	76...	1.327	0.970	0.242	4.011
17....	0.297	0.292	0.956	0.306	47....	0.820	0.731	0.682	1.072	77...	1.344	0.974	0.225	4.331
18....	0.314	0.309	0.951	0.325	48....	0.838	0.743	0.669	1.111	78...	1.361	0.978	0.208	4.705
19....	0.332	0.326	0.946	0.344	49....	0.855	0.755	0.656	1.150	79...	1.379	0.982	0.191	5.145
20....	0.349	0.342	0.940	0.364	50....	0.873	0.766	0.643	1.192	80...	1.396	0.985	0.174	5.671
21....	0.367	0.358	0.934	0.384	51....	0.890	0.777	0.629	1.235	81...	1.414	0.988	0.156	6.314
22....	0.384	0.375	0.927	0.404	52....	0.908	0.788	0.616	1.280	82...	1.431	0.990	0.139	7.115
23....	0.401	0.391	0.921	0.425	53....	0.925	0.799	0.602	1.327	83...	1.449	0.993	0.122	8.144
24....	0.419	0.407	0.914	0.445	54....	0.943	0.809	0.588	1.376	84...	1.466	0.995	0.105	9.514
25....	0.436	0.423	0.906	0.466	55....	0.960	0.819	0.574	1.428	85...	1.484	0.996	0.087	11.43
26....	0.454	0.438	0.899	0.488	56....	0.977	0.829	0.559	1.483	86...	1.501	0.998	0.070	14.30
27....	0.471	0.454	0.891	0.510	57....	0.995	0.839	0.545	1.540	87...	1.518	0.999	0.052	19.08
28....	0.489	0.470	0.883	0.532	58....	1.012	0.848	0.530	1.600	88...	1.536	0.999	0.035	28.64
29....	0.506	0.485	0.875	0.554	59....	1.030	0.857	0.515	1.664	89...	1.553	1.000	0.018	57.29
30....	0.524	0.500	0.866	0.577	60....	1.047	0.866	0.500	1.732	90...	1.571	1.000	0.000	∞

Other propositions which are more difficult to prove are

$$\sin 2A = 2 \sin A \cos A$$
$$\cos 2A = \cos^2 A - \sin^2 A$$
$$= 1 - 2 \sin^2 A = 2 \cos^2 A - 1$$

Another most important point is that the sine and cosine curves shown in Fig. 38 are merely displacements of each other. If the cosine curve is pushed to the right by 90°, or $\pi/2$ radians,† the two curves coincide. This is another property we can prove using only well-known theorems of plane geometry.

It is impossible to use these circular functions, as they are called, without making use of the number of π, which is the ratio of the length of half the circumference of a circle to its radius or the number of radians in 180°. You surely know that this number is approximately 3.141 . . . Perhaps some of you know a little about the history of its discovery. Here are some of the high points. We should occasionally consider the tremendous labors that have gone into creating what we accept as matters of course and use casually in our daily lives.

The story usually begins with Archimedes in about 240 B.C. He realized that the length of the circumference of a circle is surely greater than that of an inscribed polygon and less than that of a circumscribed polygon. In Fig. 39 is a circle with a circumscribed hexagon and an inscribed hexagon. The two lengths may easily be shown to be $6R$ and $\frac{12}{\sqrt{3}}R$.

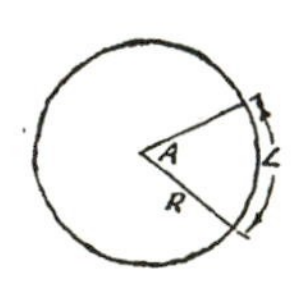

† Angles may be specified in degrees or in radians. The number of radians in angle A is the ratio of the length of the circular arc subtended by the angle to the length of the radius. In the illustration, $A = L/R$ radians.

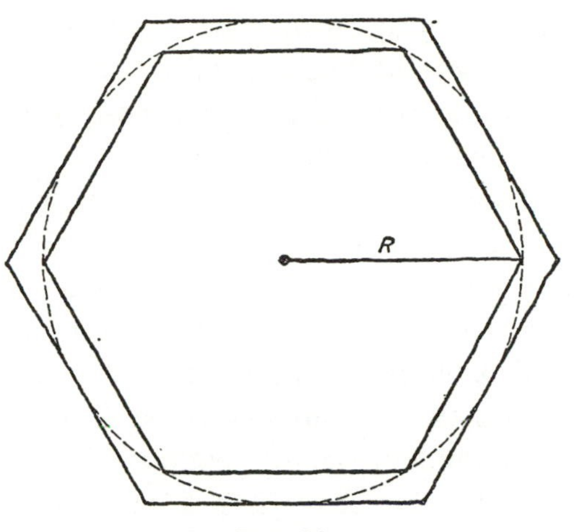

Fig. 39.

Archimedes then computed from geometrical considerations the length of 12-, 24-, 48-, and 96-sided inscribed and circumscribed polygons. Finally he concluded that the number of π lay between $\frac{223}{71}$ and $\frac{22}{7}$ or that to two decimal places the value of π is 3.14.

In the course of the ensuing centuries scholars and astronomers all over the world—in Alexandria, in China, in India—considered this problem and discovered other ways of getting at the value of π more accurately. For instance, in 1650 an English mathematician, John Wallis, discovered that

$$\frac{\pi}{2} = \frac{2 \cdot 2 \cdot 4 \cdot 4 \cdot 6 \cdot 6 \cdot 8 \ldots}{1 \cdot 3 \cdot 3 \cdot 5 \cdot 5 \cdot 7 \cdot 7 \ldots}$$

Somewhat later a Scotch mathematician, James Gregory, discovered that

$$\frac{\pi}{4} = 1 - \tfrac{1}{3} + \tfrac{1}{5} - \tfrac{1}{7} + \ldots$$

This notion of expressing a quantity, not in terms of a number specifying its magnitude, but as a series of numbers to be multiplied, divided, or added with a rule for setting up the series, is a most useful tool to which we shall come back later.

Then there are so-called experimental methods for determining the magnitude of π. In 1760 Comte de Buffon devised the following scheme. Suppose a number of parallel lines are ruled a distance a apart on a horizontal plane, and suppose that a rod of length l, which is assumed to be less than a, is dropped at random on this pattern. Buffon showed that the probability that the rod will fall across one of the lines in the plane is given by $\frac{2l}{\pi a}$. The experiment has actually been tried, the probability p has been determined, and from the ratio of $\frac{2l}{pa}$ the number of π has been found in satisfactory agreement with values established by other means.

And so the work continued. People, by one means or another, computed to more and more decimal places, made mistakes, corrected the mistakes. At present the record seems to be held by an electronic calculator which in 1949 calculated π to 2035 decimal places in about seventy hours, or about three days working night and day. It is interesting to note that the symbol π which we now use to describe the number 3.14159 . . . was so used first in 1706, and then gradually came into regular use toward the middle of the century.

And with this little digression to illustrate how apparently extraneous efforts add vitally to the evolution of this or that train of thought, we now return to our main theme—the discussion of vibrations.

A New Way of Describing Curves

The simplest kind of description of an object is a comparison with another object. *A* is like *B;* a dime is "like" the number of fingers on both hands; a photograph of John is "like" John; the plot of sin x as a function of x in Fig. 40 is "like" a plot of the vertical displacement of a spot on the circumference of a rotating wheel.

But how else *can* we describe things? The truth is that if our description of the physical world were limited to comparisons like this we could, at the very best, do no more than record what is happening at the surface. The deep-seated, underlying reasons for certain kinds of similarity would escape us. A great advance in our description comes when we go beyond a comparison of one thing with another, and consider comparisons of *different aspects of the same thing* with each other. By describing a thing or a situation in terms of itself we get at inner relationships which add a new dimension to our understanding.

This is something new—to describe a curve or shape or series of events *in terms of itself*. This is a key idea—a *turning point* in our advance into the unknown. Without this we could not have bridges, or boats, or books as we know them. We could not "see" the properties of our surroundings or "understand" the behavior of planets or atoms.

An exciting element involved in these ideas is the possibility of prediction. Actually, we are all familiar with the method to be used, but only in very simple instances. We must examine what we know carefully and discover how to use it more generally. Then we

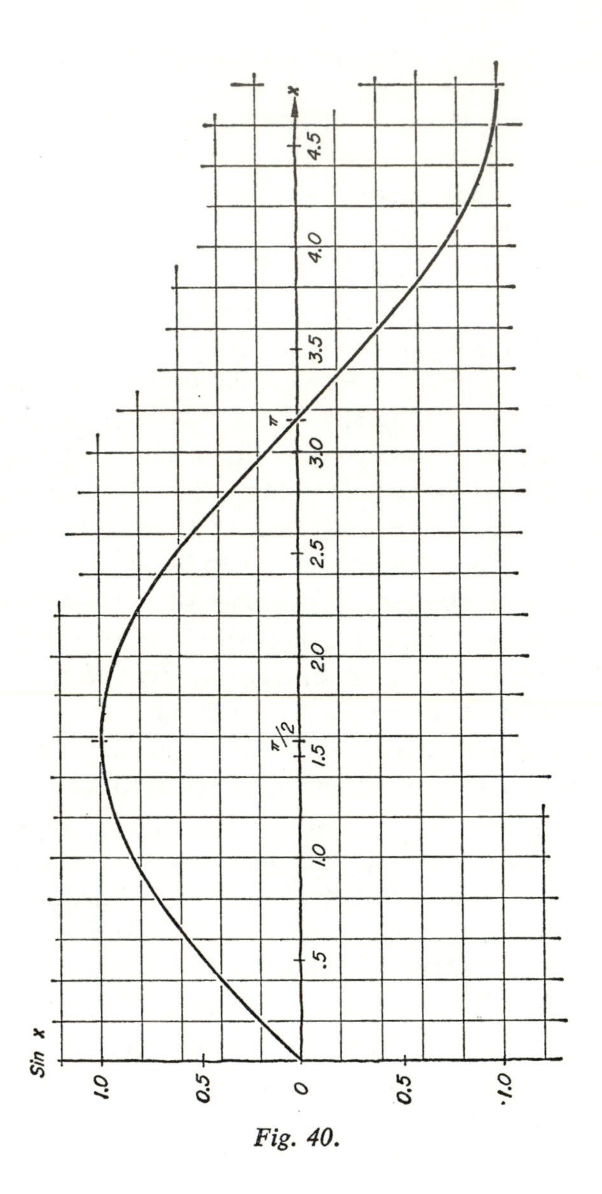

Fig. 40.

shall see how, from our ability to predict, we can formulate general propositions, new laws of nature.

Imagine that you are standing by a road, and a car is facing you a hundred yards or so away. Can you predict when it will be opposite you? Not unless you know something about its motion. If you can form an idea of how fast the car is approaching, you can tell pretty well how long it will take to reach you. A ballplayer can catch a ball because he can predict when the ball will be within his reach, and just where to put his hand at that time. These examples illustrate how we relate the position of an object to its velocity. We are able to describe a path not by a photographic trick of memory after we have seen it, but by relating two aspects of a moving object to each other, and predicting.

How about a more complicated example, say a swinging pendulum? Not only does it move, but its rate of motion is continually changing. It speeds up, slows down, stops, reverses its motion, and so on. In a qualitative way we feel that we can predict its motion, but that is not good enough. By observing the motion closely—quantitatively—we can make exact predictions. And not only that. We shall be able to predict something about its motion before it starts to move if we know how it is built, if we know its mass, and the length of its support. But this will require a series of intermediate steps.

Our discussion starts with considerations of the slope of a curve. Take, for instance, the curve plotted in Fig. 41. Here we have a function of the variable x. We call this function f, or $f(x)$, read "f of x." The value of f for any given x is the height of the curve above the x-axis. At every point this function has not only a value, but a slope like that indicated by the

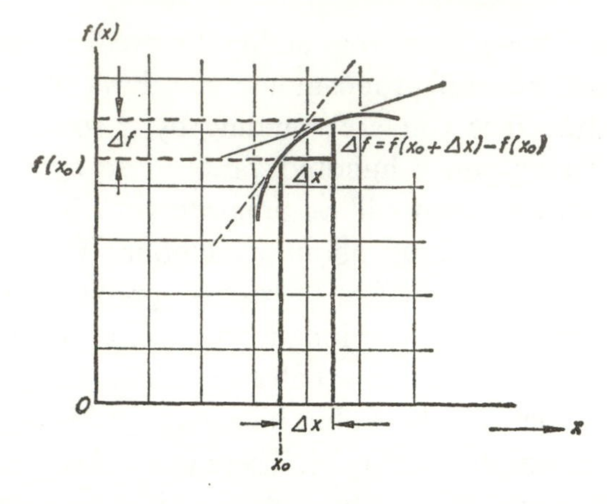

Fig. 41.

straight line through the point at x and tangent to the curve. The slope of a line with respect to a horizontal reference line we already have defined as the increase in height per unit of horizontal travel. The slope of a curve at any point is simply the slope of the tangent line at that point.

And now we need a little mathematical notation, extending and making more precise our discussion of the last chapter in connection with slopes in Fig. 31. Suppose we consider two points along the curve in Fig. 41. One is at the point x_0 and the other is at a point somewhat farther along, farther by a small amount Δx. The symbol Δ, as we said earlier, is a Greek capital delta, or D, and is used to indicate a small but finite increment. At the first of these two points the value of the function is $f(x_0)$. At the second it has increased by an amount Δf. The slope of the line through our two points is $\Delta f/\Delta x$. If we now let the interval Δx get smaller and smaller, the line through our two points will become more and more nearly coincident with the dotted tangent line. To indicate this process of having an increment Δ becoming smaller and smaller, we write d instead of Δ. Thus we have the following relations:

slope of the curve f at any point x = the slope of the tangent at that point = df/dx.

This is, I hope, an intuitively acceptable geometric relationship, and we need not go further into its significance for the purposes of this study. Actually, we are on the threshold of the differential calculus. df/dx is called the derivative of f with respect to x. The slope of a curve is "derived" from the curve itself. But you need remember only the graphic situation and understand what we mean by the italicized statement.

Let us consider the foregoing from a more concrete point of view. We have said that we could predict something about the motion of an automobile along a road if we knew its position and velocity. Let us give the axes in Fig. 41 a physical significance and see what the curve implies. Let us suppose that x represents time, and that $f(x)$ represents distance along a road. The curve then may represent the location of an automobile moving along the road. The slope of the tangent line at $x = x_0$ represents the speed of the automobile at this particular instant.

$$\frac{df}{dx} = \textit{instantaneous speed of the automobile.}$$

Note that as time goes on, or as x increases, the slope decreases. Finally the tangent is horizontal, the slope, or speed, is zero, and the automobile has come to rest. We cannot predict the motion of the automobile from its position and speed at the time x_0. The prediction would be valid for only a very short time. We must know how the speed changes.

In this case we must know how the car is decelerated when the brakes are applied.

Now let us examine this situation in another case, and just to make things different, let us consider a curve with a negative slope as in Fig. 42a. Actually

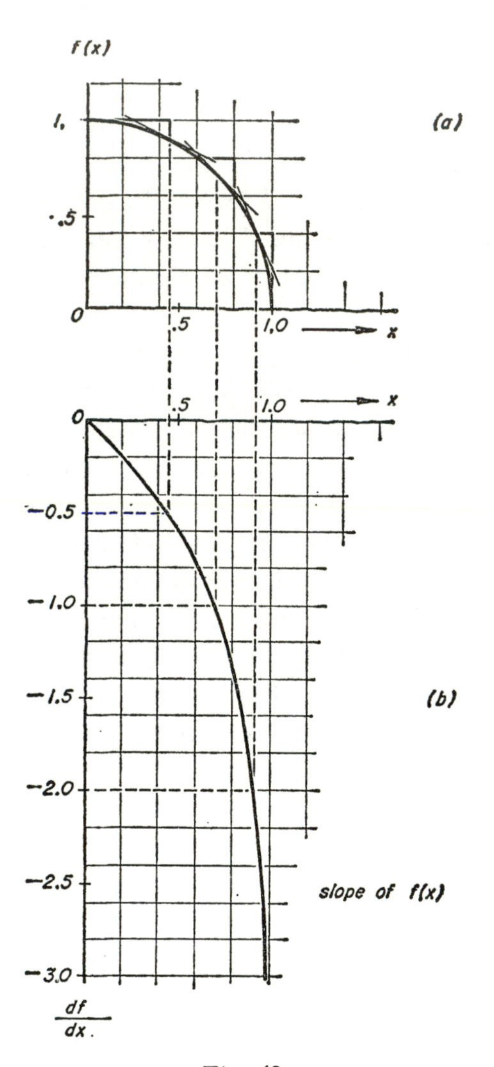

Fig. 42.

this curve is a quarter of a circle. From the curve representing the function f in Fig. 42a we can construct Fig. 42b showing the slope of f at every point. Let us call this the "slope-curve," or simply the curve showing the slope or derivative of f for any given value of x. Could we, conversely, knowing the slope of f at every point, reconstruct the function f itself? This would be a difficult task. Let us see how we might go about it.

Clearly, $f(x)$ is horizontal at $x = 0$, then slopes downward as x increases, with a rapidly increasing negative slope as x approaches 1. But where should we start the curve? We might perfectly well start it at $f(x) = 2.5$ as in Fig. 43, and then sketch in the approximately correct solid curve shown. For comparison the dotted curve, which has the correct slope

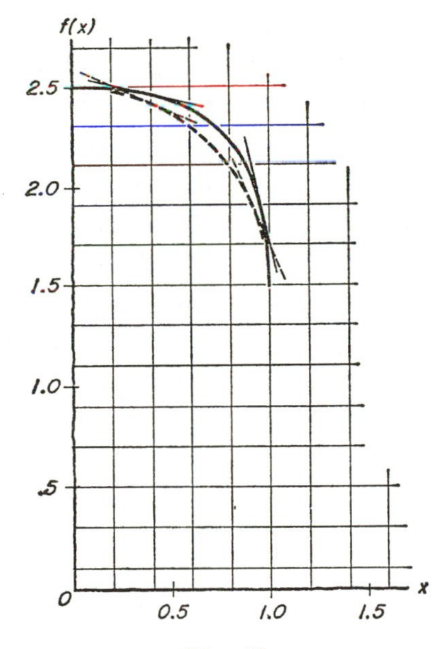

Fig. 43.

at every point, is also shown. The solid curve has too small a slope for small values of x, and too great a slope for larger values of x. By dint of gradual changes one might get the right curve, but there is nothing in the specification of df/dx in Fig. 42b to tell us where to put the curve $f(x)$ along the vertical axis.

There is this arbitrariness, that the curve which has the slope specified in Fig. 42b can be the curve shown in Fig. 41 or the same curve displaced vertically by any arbitrary amount. But if we know any one point on the curve f (for example, that at $x = 0$, $f(x) = 1$) we can construct all f unambiguously from its derivative, or slope.

We may well pause here and consider what we are achieving in this apparently abstract process of discussing the slopes of curves. We are spelling out the first few words in a new language, a language especially adapted to describing the changing universe, the rising and setting moon, the falling rain drop, the curving path of a bicycle, the flow of electricity along wires, or the sinking of a stone thrown into a pond. All life, all existence, involves change. How fast is change taking place? This, in a graphical presentation, is the slope, the *rate of change* of some measured and plotted quantity. Once we are familiar with this, we can go on to consider the relationships of different changes, of different rates of change, to each other.

The process of determining a slope, or rate of change, can be extended. If an automobile is coming to a stop, we can describe its motion by means of a curve like that in Fig. 41. Its slope at any point gives us the speed of the automobile. The application of the brake slows the car down. How can we study the way in which the driver is applying the brakes? We can do it by studying the manner in which the speed is changing. If the car is just coasting on a level road, the speed is essentially constant (except for frictional drag of one kind or another); the speed, df/dx, is constant. But if df/dx changes because of braking or acceleration, we must study the rate of change of the slope, or the slope of the curve representing speed.

We can measure the slope of a slope-curve at every point, using, for instance, Fig. 42b, and so arrive at still another curve. We must arrive at this third curve by repeating the process for arriving at the second.

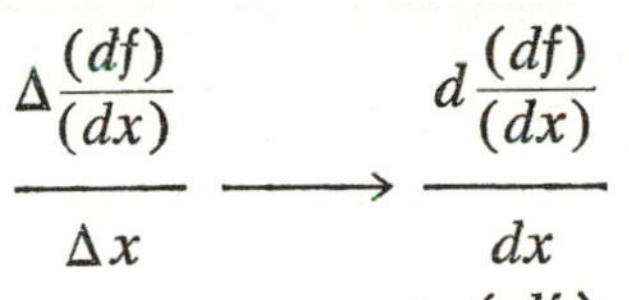

$$\frac{\Delta\frac{(df)}{(dx)}}{\Delta x} \longrightarrow \frac{d\frac{(df)}{(dx)}}{dx}$$

which we write symbolically $\frac{d}{dx}\left(\frac{df}{dx}\right)$, or $\frac{d^2f}{dx^2}$ and we call this the *second derivative* of the curve f. These three curves, and arbitrarily more still, can be derived one from the other.

Let us apply these ideas to the sin x curve, which is replotted in Fig. 44a. The sine curve starts at the origin, and its tangent at the origin slopes upward at 45°. It has a slope of $+1$, the legs of all 45° right triangles being equal. We must therefore start our plot of the slope of sin x at 1, as in Fig. 44b. The sine function reaches its maximum value when $x = 90°$, or $\pi/2$ radians. Here its slope is zero, and we must put the value of $\frac{d \sin x}{dx}$ at zero for $x = \pi/2$. And so on. For $x = \pi$, we note that $\sin x = 0$, but its slope is downward at an angle of 45°. This means

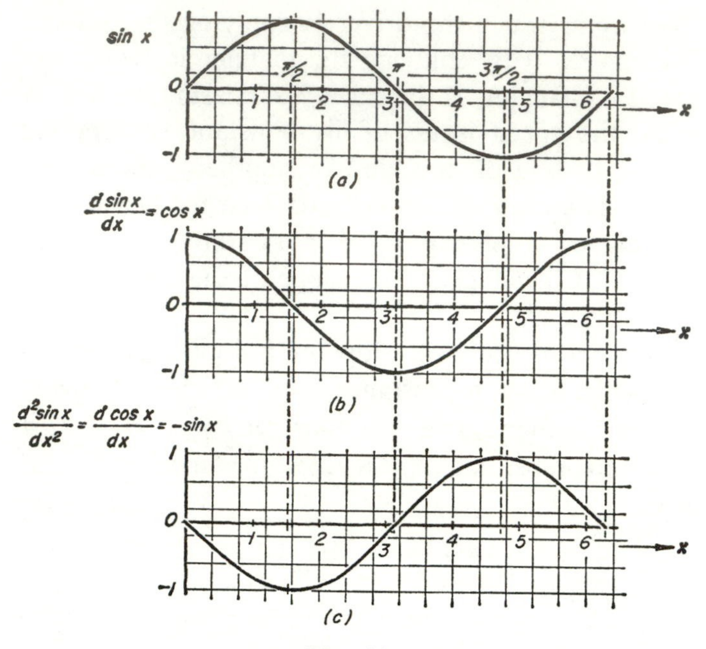

Fig. 44.

that $\dfrac{d \sin x}{dx} = -1$ for $x = \pi$. The final result is the curve shown in Fig. 44b, showing a graph of the slope of sin x, or $\dfrac{d \sin x}{dx}$ as a function of x. This is very like the original curve we started with, displaced to the left by a quarter of a cycle. But in turn a curve so displaced is cos x. You can convince yourselves that this is approximately true by careful graphical construction. Mathematically it is possible to show that this is exactly true,

$$\frac{d \sin x}{dx} = \cos x$$

We can go further, and study the slope of the cos x curve. Proceeding as above, we arrive at the curve for the second derivative of sin x, or the first derivative of cos x. It is the function $-\sin x$. Mathematically we have

$$\frac{d \cos x}{dx} = -\sin x$$

We now can state the interesting fact that for a sine function–or, for that matter, for a cosine function–the slope of the slope-curve is not identical to the original function, as you can see in Fig. 44c, but is its negative. We have

$$\frac{d^2 \sin x}{dx^2} = -\sin x$$

$$\frac{d^2 \cos x}{dx^2} = -\cos x$$

Furthermore, and this is a step which is not obvious but nevertheless true, if we are told that a function $f(x)$ has this property; namely, that

$$\frac{d^2 f(x)}{dx^2} = -f(x)$$

then we can conclude that $f(x)$ is a periodic function having the shape of a sine or cosine curve.

The properties of plane triangles you learned in Euclidean geometry may appear at first glance to exhaust the description of the relationships between the angles of a triangle and the lengths of its sides. But here we have a whole new sequence of ideas–in fact, a whole new subject, the differential calculus. These new ideas, as we shall see, shed a flood of new light on the inner nature of oscillations.

In closing this section of our discussion, let us review briefly what we have done. We have shown how to define a periodic function in terms of its derivatives. Let us illustrate how this might be done for other curves. We have seen that if a curve $f(x)$ is such that its second derivative is everywhere equal to the negative of the function itself, then the function is a periodic function. If, on the other hand, we have a curve $f(x)$ characterized by the relation

$$\frac{df(x)}{dx} = 0$$

we can conclude that $f(x)$ is a horizontal straight line, or that $f(x)$ has a constant value, for example $f(x) = 1$, as in Fig. 45a. If, on the other hand, we

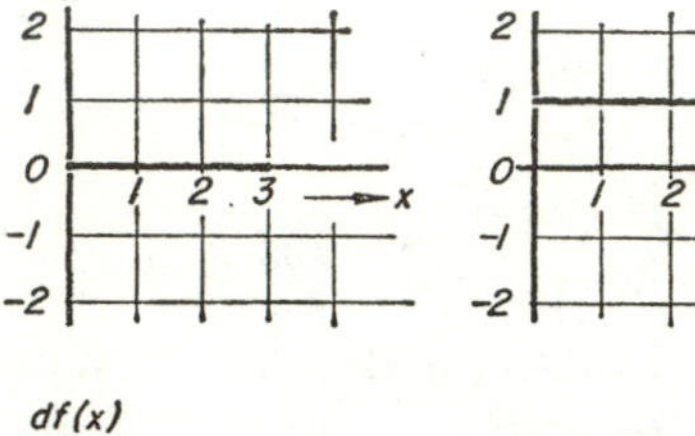

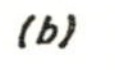

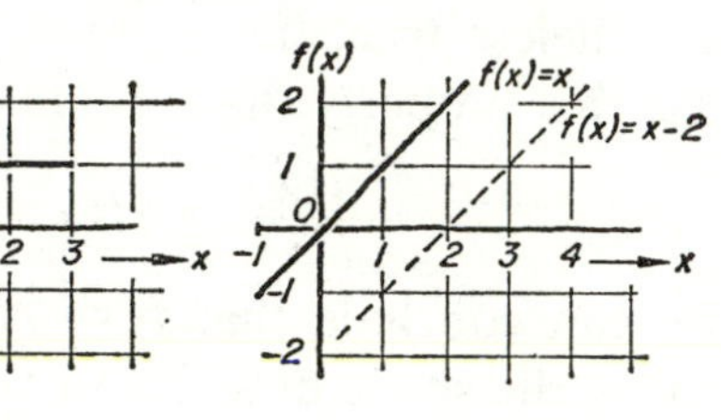

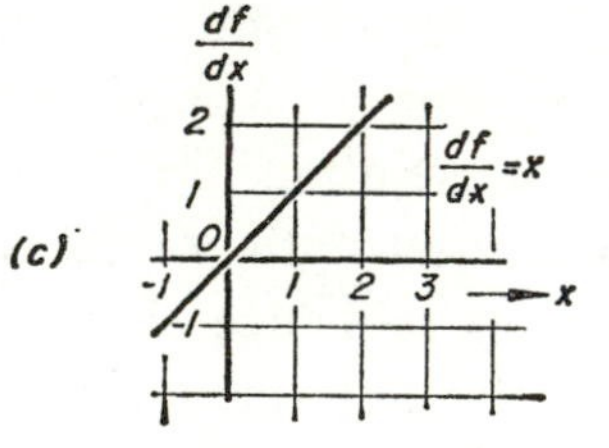

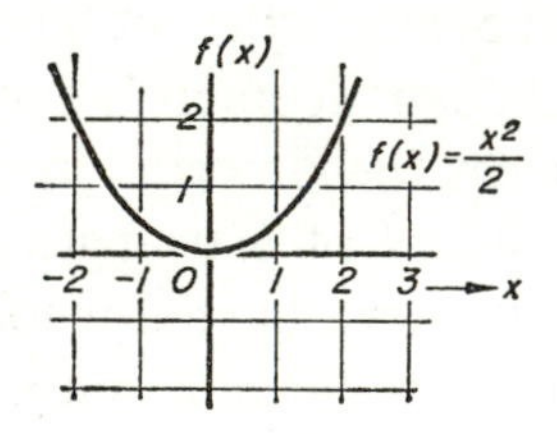

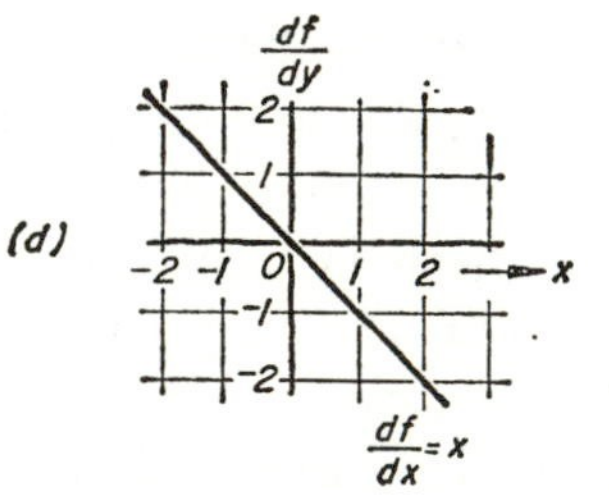

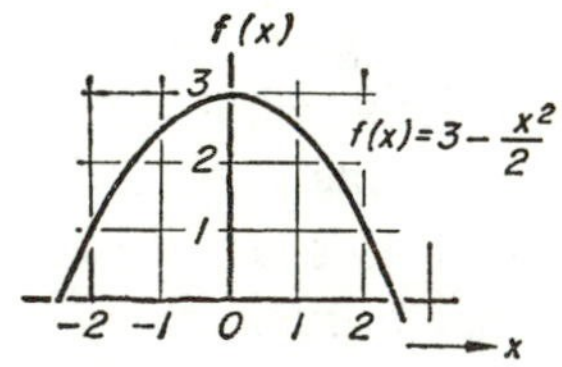

Fig. 45.

have another function characterized by the relation

$$\frac{df(x)}{dx} = 1$$

we can conclude that $f(x) = x$, or $x +$ constant, as in either case $\frac{df}{dx} = 1$. This is the equation representing a line with constant unit slope, as in Fig. 45b. Further, you can show by careful plotting that the curve

$$f(x) = \frac{x^2}{2}$$

has a slope at every point equal to x, as shown for a few points in Fig. 45c. We therefore have

$$\frac{d\left(\frac{x^2}{2}\right)}{dx} = x$$

and

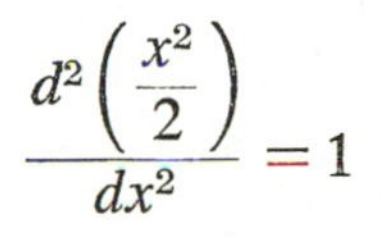

$$\frac{d^2\left(\frac{x^2}{2}\right)}{dx^2} = 1$$

In the next section we take up the reasons for the similarity between the curves we have discussed and certain kinds of physical motions. The physics, as we shall see, can be expressed very neatly in terms of the geometrical or mathematical relationships which we have brought out. If you hold this curve $f(x) = \frac{x^2}{2}$ upside down, or if you do the equivalent thing, plot $f(x) = -\frac{x^2}{2}$, as in Fig. 45d, whose slope incidentally is everywhere equal to $-x$, you may notice that the curve is somehow similar to the path of a ball thrown up into the air.

The Geometry of the Laws of Motion

The derivatives described in the preceding section have a special meaning when the graphs represent position and time in motion along a straight line. It is perhaps worthwhile at this point to discuss a very familiar problem in the language which we propose to use for describing the less familiar vibrations and oscillations. We have been discussing the shapes of arbitrary curves in terms of their slopes or first derivatives, and the slopes of curves representing slope, or the second derivatives.

At the top of Fig. 46 we have three different kinds

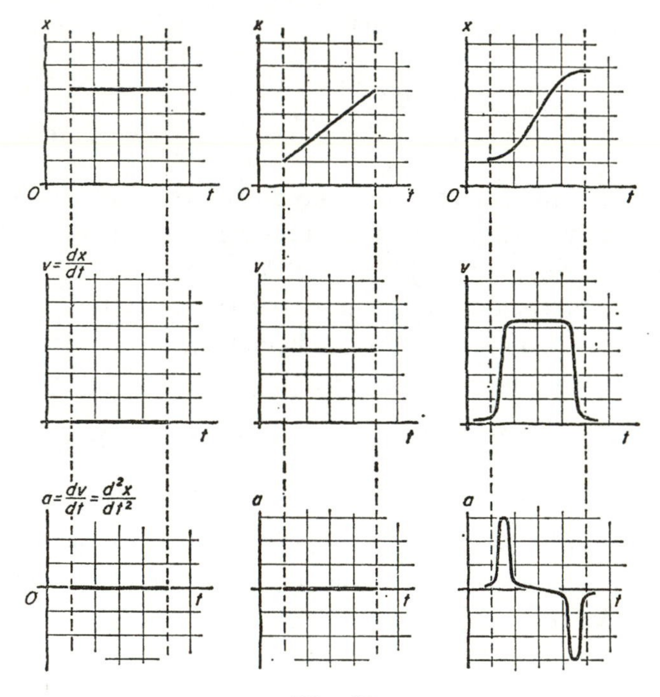

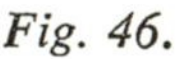
Fig. 46.

PLATE I *(Yerkes Observatory photograph)*

PLATE II *(Harvard College Observatory photograph)*

PLATE III

Molto vivace. (♩. = 116.)
2 Flöten.
2 Hoboen.
2 Klarinetten in C.
2 Fagotte.
2 Hörner in D.
2 Hörner in B.
2 Trompeten in D.
Alt- u. Ten-
Posaunen.
Baß-
Pauken in
1. Violinen.
2. Violinen.
Bratschen.
Violoncelli
u. Kontrabässe.
zu 2

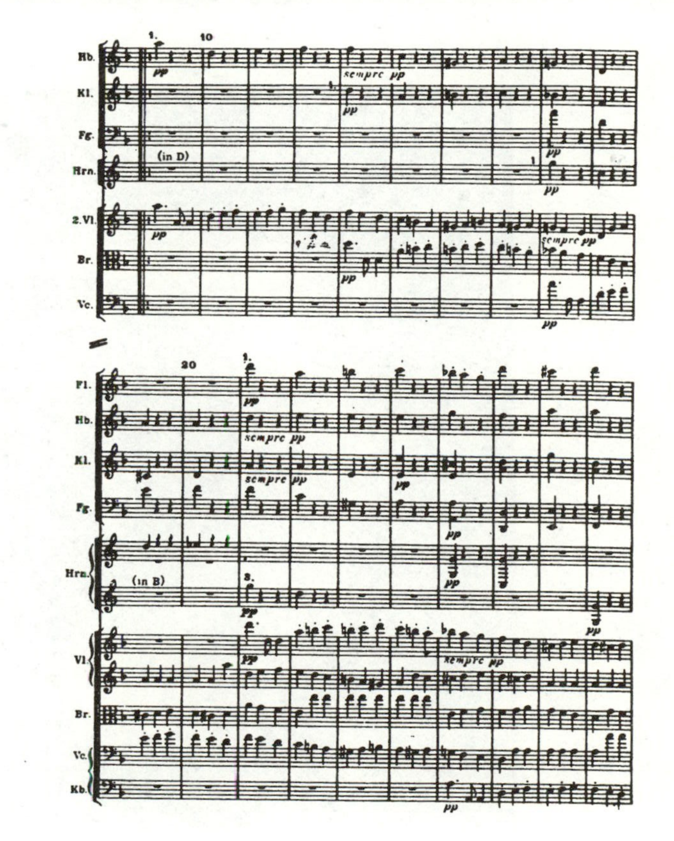
Hb.
Kl.
Fg.
Hrn.
(in D)
2. Vl.
Br.
Vc.
sempre pp
Fl.
Hb.
Kl.
Fg.
Hrn.
(in B)
Vl.
Br.
Vc.
Kb.
sempre pp

PLATE IV

PLATE V *(Mount Wilson and Palomar Observatories)*

of curves. The first is a horizontal straight line, the second is an inclined straight line, and the third is a curved line. Mathematically, if we let the ordinate represent distance along a path and the abscissa represent time, then the first curve in which the distance remains unchanged as time moves on represents a point at rest. The second curve in which the distance x increases at a uniform rate as time increases represents an object in uniform motion. Its speed is the slope of this line. The third curve represents an object starting from rest and moving, in the interval of time indicated, from one point to another point farther along the x-axis and then coming to rest at this farther point. In this last curve, as time increases, the slope or velocity is at first zero but gradually increases as the particle speeds up; then the slope decreases until, finally, the particle comes to rest in the new position.

We can plot the slopes of these curves, or the corresponding velocity of the particle, as a function of time, as has been done in the second series of curves in Fig. 46. In the first graph the velocity is everywhere zero; in the second the velocity is constant in the interval of time represented; in the third the velocity starts at zero, increases, reaches a maximum value, and then decreases to zero again.

Finally, we can represent the accelerations of the particle in terms of the slope of a velocity curve. These are particularly simple. In the first two graphs the accelerations are zero since the velocity remains constant, in the first case at the value zero and in the second at a finite value. In the third case the acceleration plot is more complicated. While the particle is at rest the velocity is constant, and the acceleration is therefore zero. As the particle begins to move, the acceleration increases until the slope of the velocity curve has a maximum value and then decreases again to zero during the interval of time in which the particle has a constant velocity. Finally, when the particle is slowed down, the acceleration becomes negative since the velocity is decreasing. It goes through a maximum negative value and finally returns to zero again when the particle is brought to rest.

These purely geometrical relationships have a physical interpretation. They describe the physical laws governing the motion of particles in particular situations. A fundamental law of mechanics, first enunciated by Newton, can be expressed in the form

$$F = ma$$

In words the equation says that in order to change the state of motion of a mass m, a force F is required, such that it will equal the product of mass and the rate of change of velocity, or the acceleration. Consider the application of this law to the motions described in Fig. 46. In the first two instances, where the velocities were respectively zero and constant with a finite value, there were no accelerations, and consequently no forces. Only in the third case would forces have to be applied to change the state of motion, first from rest to motion with a finite velocity, and then to reduce this velocity to zero again. The last curve on the lower right-hand side of Fig. 46 is a representation not only of the acceleration of the particle but of the force required to produce this acceleration.

We now proceed one step further to consider the motion of a particle in the earth's gravitational field near the surface of the earth, where we may assume the acceleration due to gravity (g) to be constant. If we let x represent the height of the particle above the earth and t the time and consider the downward pull of gravity to be in the direction of the negative x axis, then Newton's law of motion may be put into the form

$$F = ma = -mg$$

And now, making use of the fact that acceleration may be written as the second derivative of the height of the particle above the earth, or d^2x/dt^2, we are led to the expression which governs the motion of a particle being acted on by the earth's gravitational field as

$$F = -mg = m\frac{d^2x}{dt^2};$$

$$\frac{d^2x}{dt^2} = -g$$

This is the mathematical description of a physical law. From it we can deduce the vertical motion of any particle in the earth's field of force. This procedure is outlined in Fig. 47. Remember that the acceleration is the slope of the velocity curve. We can construct the velocity at any point according to the equation

$$\frac{dv}{dt} = -g$$

This equation tells us only how the velocity changes. It describes motions such that velocities change by the same amount in successive time intervals, for example, 10 mph, 9 mph, 8 mph, 7 mph as measured now, t secs from now, $2t$ secs from now, $3t$ secs from now, and so on. The acceleration is constant when the velocity change is constant. In order to apply it we must stipulate the magnitude of the veloc-

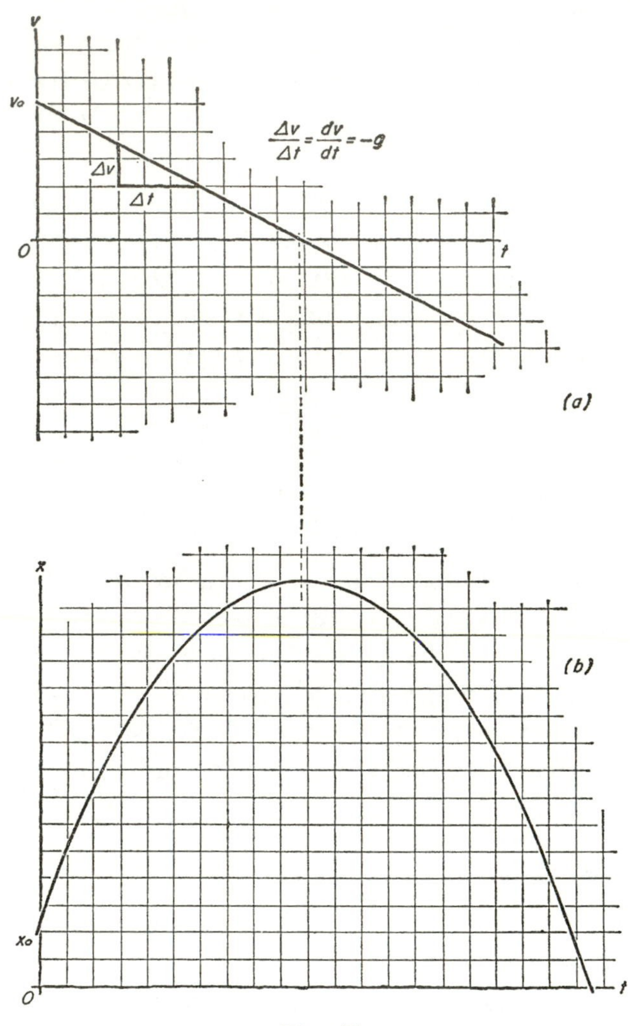

Fig. 47.

ity at some initial point. Obviously the trajectory of a baseball will be different if it is dropped, or if it is thrown up into the air at some initial velocity. In the upper curve of Fig. 47 we show a curve of velocity as a function of time. Here we assume that the velocity at some initial time had the arbitrary value v_0. The slope of this curve must be $-g$, leading to the straight line indicated. The velocity decreases, becomes zero, and then becomes negative. This obviously must represent a particle, for example a baseball, which initially is thrown upward with a positive velocity, comes to rest at the point $v = 0$, and then becomes negative as it falls. For example, if the initial upward velocity were 10 m/sec, the acceleration of gravity, which is 9.8 m/sec^2 in a downward direction, would reduce the velocity to 0.2 m/sec at the end of the first second, then to −9.6 m/sec (in a downward direction) at the end of the second second, to −19.4 m/sec at the end of the third second, and so on. The time at which the particle reaches its greatest height is the time at which it comes to rest, or

$$1+\frac{0.2}{9.8}=1.0204 \text{ secs}$$

The equation for the velocity curve is

$$v=v_0-gt$$

or in the example discussed

$$v = 10 - 9.8t$$

From the velocity curve we can derive the position as a function of time. This is not as easy as the reconstruction of the velocity, but can be done with the help of the reasoning at the end of the previous section. In principle it involves merely filling in the following table:

Time	*Velocity*	*Distance Moved in 1 sec*
0	10 m/sec	———
1 sec	0.2 m/sec	———
2 sec	– 9.6 m/sec	———
3 sec	–19.4 m/sec	———

This will give an approximate answer only, because the velocity is not actually constant during a second. A better result is obtained using smaller time intervals. The correct result, using arbitrarily small time intervals, is illustrated in Fig. 47b. Here again there is an unknown. Since the velocity curve that we have constructed merely gives us the change of position with time, we must specify where the particle is at some initial time. But, having done this, we can construct a curve whose slope at every point is just that given by the previously constructed velocity curve and which passes through the specified initial point.

THE GEOMETRY OF OSCILLATORY MOTION

If you have followed these discussions, which are really much more formal and mathematical than physical, you should have no trouble with this next one, in which we make a decided jump. We shall progress from rules and grammar, from saying rather dull things like, "The black cat chases the white dog," to the more interesting phase of using language to tell a new story.

In describing oscillatory motions and relating the properties of sine curves to physical laws, we find that we must turn a corner, a rather unexpected corner, beyond which we see an unexpected view. In a way, we can anticipate this view.

The motions of many, many different oscillating systems can be represented with sine curves. What are the characteristics of these oscillating systems? Just as we have done for the falling body, we may expect in describing any particular oscillation to have to include a specification of initial conditions. In describing the vibration of a tuning fork we must specify for some initial time the position and velocity of the vibrating member. These two quantities determine, first of all, what is called the "phase" of the motion–that is, the particular instant at which the vibrating system goes through its equilibrium position; and, further, establish the amplitude of the motion–that is, the extent of the vibration. A tuning fork, or any other system, can be set into vibration with a small amplitude or with a large amplitude, depending on how hard you strike it or pluck it.

But there is one aspect of vibrational motion that is not dependent on the initial conditions, and is yet very different for different vibrating systems. What determines the pitch of the tuning fork? Any one tuning fork always vibrates with the same frequency, or at the same pitch, no matter how it is struck. The parts go through the same number of cycles per second. This frequency is not dependent on the initial conditions and thus it differs from the amplitude. It is something inherent, a property of the vibrating system itself. It must somehow come out of Newton's laws of motion and the properties of the tuning fork.

The analysis that reveals this inherent something is the unexpected corner I was speaking of. But before we get too deep in the analysis, let us summarize what we have already found out. If y is an angle expressed in radians, then $\sin y$ is a curve which we know how to construct. It is drawn in Fig. 48. We

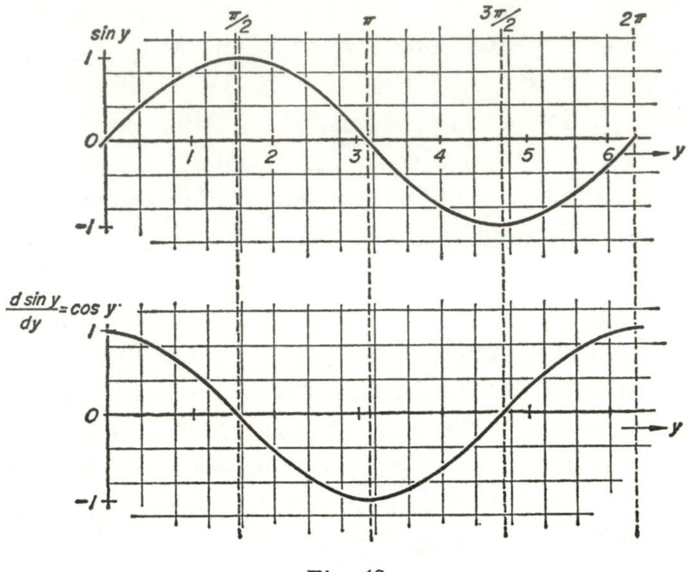

Fig. 48.

also know the derivatives of this function

$$\frac{d \sin y}{dy} = \cos y$$

$$\frac{d^2 \sin y}{dy^2} = \frac{d \cos y}{dy} = -\sin y$$

These are formal, geometrical or mathematical relationships. If, however, we come to the description of a vibrating system, for example the weight on a spring in Fig. 49, we must consider a displacement x measured in some units of length as a function of time t measured in seconds. The physical situation might be described as in Fig. 50. Here the amplitude A is 3 inches and the period T is 1.5 seconds. The slope of this curve has a physical meaning. The derivative $\frac{dx}{dt}$ is a velocity, and the slope of this velocity curve is an acceleration, and

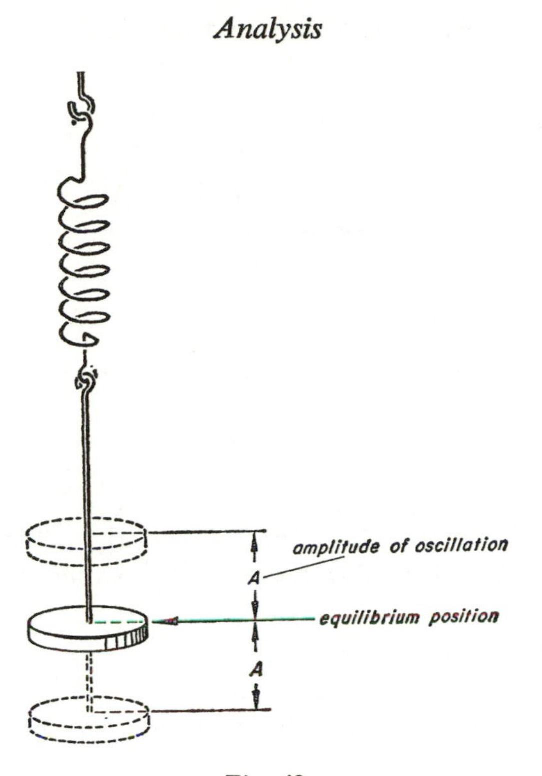

Fig. 49.

the acceleration of a particle is related to the force acting on it. Notice that we could not specify the curve in Fig. 50 by writing simply $x = \sin t$. Such a curve would have the wrong amplitude and period. In order to describe the curve in Fig. 50 we must introduce suitable scale factors for both the displacement and the time. We can specify the amplitude A in inches and the period T of the motion in seconds by writing

$$x = A \sin 2\pi \frac{t}{T}$$

Whenever the sine has its extreme value ± 1, the displacement x of the particle is $\pm A$ inches. Thus,

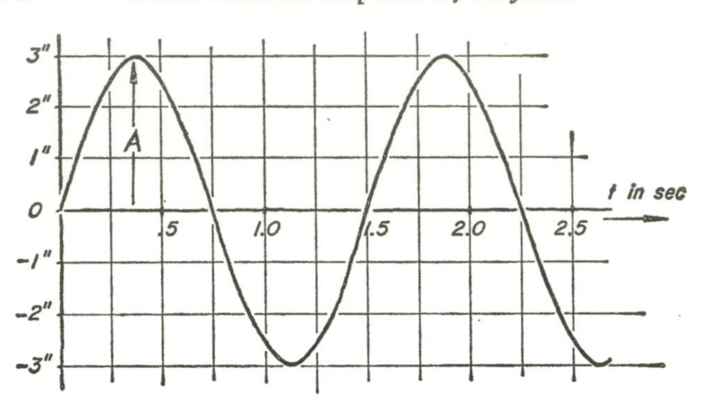

Fig. 50.

for example, if $t = \frac{T}{4}$, then $\sin 2\pi \frac{t}{4t} = \sin \frac{\pi}{2} = \sin 90° = 1$ and the displacement at this time will have its maximum possible value, namely A. If we are describing an oscillation having an amplitude of 3 inches, we must put $A = 3$ inches. Further, whenever the time changes by some integral number of times the period T, then the number $2\pi \frac{t}{T}$ increases by some integral multiple of 2π, and the sine function itself remains unchanged. Thus in Fig. 50 we find that when $t = 0.5$ secs the displacement is about 2.75 inches, and decreasing. The period being 1.5 secs, we expect to find that at times one period later, or $0.5 + 1.5 = 2$ secs, or 2 periods later, $0.5 + 3 = 3.5$ secs, or 3 or more periods later, the displacement will again be 2.75 inches and decreasing, as is actually the case.

Now, however, we are faced with a new problem. We have studied the process of differentiating the function $\sin y$ with respect to y, but now we must consider how to differentiate the displacement x in inches with respect to a time t in seconds in order to arrive at a physical velocity. This we can do. By dividing the above equation by A we arrive at the expression

$$\frac{x}{A} = \sin 2\pi \frac{t}{T}$$

Let us now define the following numbers $x^* = \frac{x}{A}$ and $t^* = \frac{2\pi t}{T}$. We may write a description of our sinusoidal motion in terms of x^* and t^*.

$$x^* = \sin t^*$$

The slope, or derivative, of this function we know.

$$\frac{dx^*}{dt^*} = \cos t^*$$

The second derivative is

$$\frac{d^2x^*}{dt^{*2}} = -\sin t^*$$

This seemingly arbitrary mathematical procedure has a simple physical explanation. An Englishman wants to buy a suit at Macy's. The price is marked as $34.95. That may seem clear to you, if you are used to American currency, but the Englishman wants to compare this with a suit costing £10 in London. So he must find the scale factor A = number of dollars per pound. The price of the suit in pounds sterling is then $\frac{\$34.95}{A}$. To him the pound is a "natural," or especially significant unit. In our spring problem, the displacement which an observer measures may be given as 2.75 inches, but the inner relationships of the spring problem are greatly simplified by describing the displacement in terms of another unit, namely the amplitude A. Thus if the displacement in inches is x, the "natural" or especially significant unit from the point of view of the spring is $x^* = \frac{x}{A}$. This is obvious. What is perhaps not so obvious is that if the price per suit at Macy's goes up by $1 per suit, the price change in pounds sterling is $\frac{\$1}{A}$. Or going back to the spring, if the displacement changes by an amount dx, then the change in the "natural" units is

$$dx^* = \frac{dx}{A}$$

Similarly, we find that

$$dt^* = \frac{2\pi dt}{T}$$

and therefore that

$$\frac{dx^*}{dt^*} = \frac{\left(\frac{1}{A}dx\right)}{\left(\frac{2\pi}{T}dt\right)}$$

$$= \frac{T}{2\pi A}\frac{dx}{dt}$$

We can also multiply through by $\frac{2\pi A}{T}$, and get

$$\frac{dx}{dt} = \frac{2\pi A}{T}\frac{dx^*}{dt^*}$$

This leads to the expression we want for the velocity of a particle whose position is given by

$$x = A \sin 2\pi\frac{t}{T}$$

$$\frac{dx}{dt} = \frac{2\pi A}{T}\frac{dx^*}{dt^*} = \frac{2\pi A}{T}\cos t^* = \frac{2\pi}{T}A\cos 2\pi\frac{t}{T}$$

This gives us the velocity of the particle in inches per second in terms of the time t in seconds. Similarly we get the acceleration in inches per second per second by applying the scale factors as required.

$$\frac{d^2x^*}{dt^{*2}} = \frac{d}{dt^*}\frac{(dx^*)}{(dt^*)} = \frac{d}{\frac{2\pi}{T}dt}\frac{\left(\frac{1}{A}dx\right)}{\left(\frac{2\pi}{T}dt\right)}$$

$$= \frac{1}{A}\left(\frac{T}{2\pi}\right)^2 \frac{d^2x}{dt^2}$$

$$= -\sin t^*$$

The last two lines state that

$$\frac{1}{A}\left(\frac{T}{2\pi}\right)^2 \frac{d^2x}{dt^2} = -\sin t^* = -\sin 2\pi\frac{t}{T}$$

$$\left(\frac{T}{2\pi}\right)^2 \frac{d^2x}{dt^2} = -A \sin 2\pi\frac{t}{T} = -x$$

This last bit of algebra yields the result we have been seeking.

$$\boxed{\frac{d^2x}{dt^2} = -\frac{4\pi^2}{T^2}x}$$

Let me say again that what we have done here is to indicate in a way which may be physically or intuitively plausible to you something so important that before we finally accept the conclusions, the reasoning must be examined in detail. This mathematicians have done, and the result can be put on a very firm basis. I have always found, however, that it is extremely difficult to appreciate the significance of a long, detailed sequence of rigorous arguments the very first time that you hear them. It has always been very helpful to me in studying to go through a simpler plausible analogous argument before tackling a more rigorous one. This approach is often useful in experimenting. If we wish to make a new and difficult measurement, which in its final form requires elaborate apparatus, it is usually desirable first to make a cruder measurement in a simple way to find out approximately how things are going to turn out. Our simple plausible arguments about oscillations lead us to the following important conclusion in a very few steps.

According to Newton's law of motion we may write

$$F = ma = m\frac{d^2x}{dt^2}$$

For a body moving with simple harmonic motion

$$x = A \sin 2\pi\frac{t}{T}$$

We have derived the expression,

$$\frac{d^2x}{dt^2} = -\frac{4\pi^2}{T^2}x$$

Combining these, we get

$$F = m\frac{d^2x}{dt^2} = -m\left(\frac{2\pi}{T}\right)^2 x$$

or

$$F = -kx \text{ where } k \text{ is a constant. } k = m\left(\frac{2\pi}{T}\right)^2$$

Determination of a Period of an Oscillation

It may take a while to digest the full significance of what we have written down. Let us see how we can say in words what is implied in the above symbols. First of all, we are dealing with a system in which there is a force opposing any displacement. This is a restoring force, $F = -kx$, whose magnitude is proportional to the magnitude of the displacement from equilibrium and directed toward an equilibrium position. Further, if a disturbing force is applied and then withdrawn, the resulting motion will be oscillatory and will be described in detail by a circular function such as the sine. We conclude this because, if $F = -kx$, then according to Newton's laws, there must result an acceleration

$$\boxed{\frac{d^2x}{dt^2} = a = -\frac{k}{m}x}$$

and we know from our geometrical studies that on an x-t plot we can describe such an acceleration in terms of a periodic function–whose period is related to $\frac{k}{m}$. In fact

$$\left(\frac{2\pi}{T}\right)^2 = \frac{k}{m}$$

or

$$\boxed{T = 2\pi\sqrt{\frac{m}{k}}}$$

where m is the oscillating mass and k is the force constant, or the force per unit displacement of the elastically held mass. For example, the weight on a spring illustrated in Fig. 49 may be displaced from its equilibrium by an upward or a downward force, and the amount of this displacement will be proportional to this force. The spring constant k is the force per unit displacement. In a stiff spring k is large; in a very flexible spring k is small. Furthermore, we should expect that if a weight hanging on the bottom of a spring is disturbed and then released, it will oscillate up and down. The amplitude of the motion will depend on the violence of the disturbance.

We have just turned the corner I spoke of. We are in a position as the result of this analysis to understand what determines the period of an oscillation. It may seem plausible, or even intuitively obvious, to some of you that the stiffer the spring the shorter the period and the greater the mass the longer the period. But even if you knew the mass on the end of a spring and could measure the force constant, such intuitive knowledge would not enable you to specify just what the resulting period would be in seconds. From the equations, however, we can see what the period must be. It must be 2π times the *square root of the quantity: mass divided by the force constant.*

This is a degree of true and precise understanding that constitutes one of the great beauties and fascinations of physics.

This analysis is applicable, in one way or another, to all vibrating systems. We shall apply it, by way of illustration, to one more, the pendulum. What determines the period of a pendulum? How fast, for example, would a 1 kilogram mass on the end of a string 1 meter long oscillate? To settle this, we need only to calculate the force constant of the pendulum, or the horizontal force required to produce unit displacement of the mass m of the pendulum bob. If this force is just k times the displacement, we would expect the pendulum to oscillate with a period $T = 2\pi\sqrt{\frac{m}{k}}$, and moreover that this period would be independent of the amplitude of the swing.

From the illustration in Fig. 51 we see that the

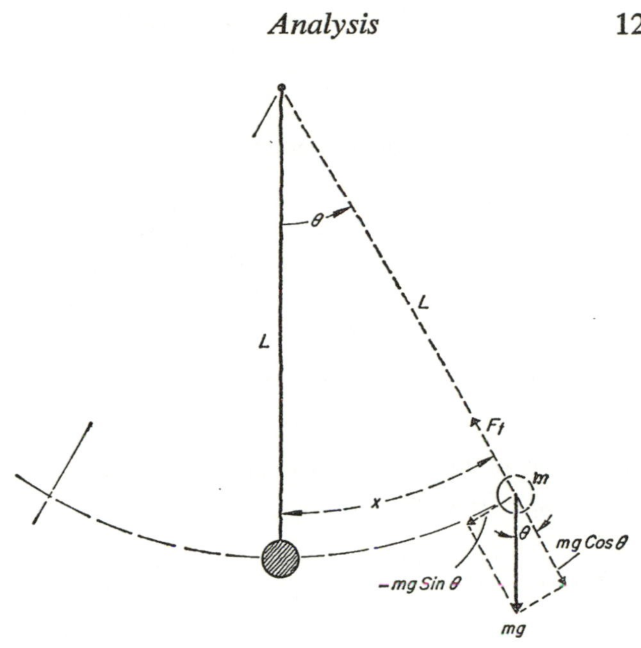

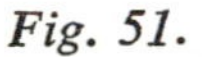
Fig. 51.

mass m has two forces acting on it—the force of gravity mg acting downward, and the tension F_T of the string. Forces can be resolved into components. The downward force mg is equivalent to two other forces at right angles to each other. One has the magnitude $mg \cos \theta$ and acts in the direction of the pendulum string. It exactly cancels the force F_T exerted by the string. The other component, $mg \sin \theta$, is uncompensated, and is the restoring force acting on the bob of the pendulum to restore it to its equilibrium position. Since it acts in such a direction as to decrease the displacement θ, it is written

$$-mg \sin \theta$$

The displacement x along the arc is related to the angle θ as follows

$$\theta = \frac{x}{L}$$

and our basic equation becomes

$$F = ma = m\frac{d^2x}{dt^2} = -mg \sin\frac{x}{L}$$

This is not the anticipated form. But if we notice that for small values of the angle the plot of the sine of an angle as a function of angle is a *straight line having unit slope,* we see that for small values of y

$$\sin y = y$$

The degree of approximation involved here is illustrated in the following table.

Angle in Degrees	*Angle in Radians to 4 decimal places*	*Sine of the Angle to 4 decimal places*
0	0.0000	0.0000
1°	0.0175	0.0175
2°	0.0349	0.0349
3°	0.0524	0.0523
4°	0.0698	0.0698
5°	0.0873	0.0872
10°	0.1745	0.1736
15°	0.2618	0.2588
20°	0.3491	0.3420

For small displacements we may write

$$\sin\frac{x}{L} = \frac{x}{L}$$

Then we find that the equation describing the motion of the pendulum does have the form required for periodic motion.

$$m\frac{d^2x}{dt^2} = -\frac{mg}{L}x$$

The force constant k is $-\frac{mg}{L}$, and the period of the motion is

$$T = 2\pi\sqrt{\frac{m}{k}} = 2\pi\sqrt{\frac{L}{g}}$$

The mass cancels out of the expression. For a mass concentrated at the end of a string of negligible mass, the period depends on the length of the string alone. For a string 1 meter long at the surface of the earth where $g = 9.8\ m/secs$ the period is

$$T = 2\pi\sqrt{\frac{1}{9.8}} = 1.99 \text{ secs}$$

This long series of arguments is like a graceful arch relating detailed specific facts about the world on the one hand to the broad, generally applicable abstract laws of physics on the other. There is a similarity between the inspired architectural arch in the church with which we started this discussion, and the arch of ideas which we have traced out. We could carry the analogy further, comparing the sketches and drawings, the experience and testing, the scaffolding and supports that went into the making of this arch, with the background of theory and experiment and observation that went into our result. And when the scaffolding is removed, and the congregation comes in to use the building which this arch supports, they see the result in all its splendor and simplicity.

$$F = -kx$$

$$T = 2\pi\sqrt{\frac{m}{k}}$$

There is more here than just formulas, when your inner eye can see it.

CHAPTER 4

Experimentation

This year, next year, every year, scientists will put into operation new sets of instruments for examining the physical world. Some of the devices will take man in directions he already has traveled—but a little further than he so far has managed to go. But others are designed for wholly new explorations—into the corners of the universe, the nooks and crannies that always have been dark and inaccessible. On these adventurous explorations science will set out simply to "see what we shall see."

Columbus sailed away to find a new trade route to the "Indies." He found the "Indies," sure enough, but not quite as he expected. There were two huge continents in the way. The ships which Spain provided, which Queen Isabella turned over to Columbus and his crew, made possible these eventful and fateful journeys. They made it possible for venturesome men to do in fact what they had dreamed about for many years–to do what they were powerfully drawn to do.

Similarly today, many new scientific investigations are being made possible by national governments. A new magnet laboratory is currently being turned over to a group of us at M.I.T. to operate so that scientists from all over the United States, actually from all over the world, can carry out in fact projects that, without the new facilities, would remain impracticable dreams. In this last chapter we shall take a look at problems related to the building of this magnet laboratory and its magnets. All too familiar are newspaper and magazine stories about the unveiling of new laboratories that can perform feats that used to be impossible. Such accounts often stick to superficial descriptions of the shiny new twentieth-century palaces where seeming miracles occur over and over. These accounts soon lose interest. But the real story of Columbus' ships is fascinating still, though transoceanic travel by sea or by air is undertaken almost as casually as a weekend in the country. Perhaps the story of magnet design will be interesting even when the strong magnets which are now being assembled for the first time are commonplace.

Experiments inside matter are going on all the time. Experimenters direct X-rays into solids and liquids and study the manner in which the rays are deflected to learn something of how the atoms are arranged. High-energy particles produced by accelerators not only penetrate matter but proceed into the atoms and on into the atomic nuclei, and break them apart. The fragments which we find then tell us about the structure of the bombarded material. Magnetic fields also penetrate into matter. Their effects are more subtle than the effects of the rays we send in as probes. The fields disturb, usually only very slightly, the charged and magnetic particles of which nuclei, atoms, solids, liquids, and gases are composed. These slight effects can be observed very precisely.

Magnetic fields in matter are like very gentle, but firm and sensitive hands, that can probe into every single nook and cranny. It has occurred to me, while a doctor poked around at my abdomen to make sure that this place isn't swollen or that part infected, that magnetic fields are much more effective in probing into solids than his hands were in probing my insides. But even the doctor's probings seem very delicate and effective compared to methods to which we sometimes have resorted. Putting an atom into a gas discharge tube, a fluorescent lamp for instance, and hitting it with electrons to make it radiate its characteristic spectrum of colored lights, has been likened to throwing pianos downstairs and analyzing the sounds they emit to discover how they are built. A magnetic field reaches into the atom and distorts it slightly. It changes the pitch of the radiation. It gives the experimenter the services of an expert piano tuner.

Inside solids there are recognizable clouds of electrons around individual atoms, but in the outer regions, where atoms come into contact with each other, there is likely to be a widespread sea of electrons that is more truly a part of the entire solid, or a large region in it, than a part of any one atom. This sea of electrons, although supported by the atoms, has characteristics of its own. It is responsible for many of the properties of solids that we observe and use. It is full of crisscrossing waves. Islands in this sea, the result perhaps of cracks and impurities, alter its behavior and properties. Heat can produce flows. Altering the chemical composition can produce high and low tides. Magnetic fields can reach down into it, push it aside, produce local eddies, and retard its flow.

Way down deep inside each atom there is the isolated atomic nucleus. It is surrounded with spherically symmetrical shells of electrons, and the nucleus and its surrounding electrons form a fluid which ceaselessly jiggles about. The amount of this jiggling depends on the temperature of the solid or liquid of which it is a part. This motion is in great contrast with the angular orientation of a nucleus. The orientation is extraordinarily fixed. In respect to its orientation the nucleus is extremely isolated. This is most important, it turns out, for our study of solids. To give you some idea of the order of magnitude of the quantities in which we are concerned, we might point out that the jiggling to and fro of nuclei in solids is of the order of 10 to the power 14 times per second, or a hundred million million times between two ticks of a clock. On the other hand, the times between reorientations may be as much as many seconds. This period might remind us, on a different scale, of the stability of a clock whose pendulum bobs to and fro once a second. The reorientation of the pendulum if it were on the same scale as the atomic nucleus would occur only once in about a million years. That is roughly the time that it takes a pendulum going to and fro once a second to perform a hundred million million oscillations.

The explanation goes back to Newton's laws of motion. The law says that for every action there must be an equal and opposite reaction. The nucleus has practically nothing to get hold of. In order to turn around or change one's position in a chair it is necessary to take hold of the sides, or push against the floor with one's feet. One must push against something. The nucleus in its spherical bed is unable to interact in a way to reorient it with respect to its surroundings except by means of magnetic fields or certain special kinds of unsymmetrical electrical fields. Through these fields, however, it can occasionally experience the kind of a twisting push necessary to face in a new direction. Magnetic fields can apply torques, as the reorienting pushes are called, to atomic nuclei and make them precess, just as the torque exerted by a gravitational field can make a top precess. By "tuning in" on this precession and seeing how long it takes for it to be modified we are able to find out a great deal about the environment of the atomic nuclei in a solid.

Experiments on the internal constitution of solids based on the phenomena that we have described are going on in scientific laboratories all over the world, all the time. But, just as microscopes with greater resolving power are wanted by microscopists, just as hotter furnaces with well-controlled atmospheres are wanted all the time by metallurgists, just as beams of particles with higher energies and greater control are wanted by nuclear physicists, so stronger and better control of magnetic fields is wanted by solid state physicists. We shall try to give you here some idea of the kind of work, the kind of analysis, the kind of effort needed in order to produce better magnetic fields, which in turn can take us into new areas of experimentation, and make it possible for us to see new things in the structure of matter.

Magnetic Fields

Magnetic fields are commonly measured in "gauss." While the gauss is an arbitrary unit, and alternatives exist, it is probably more widely accepted than shillings, pounds, and pence, or kilometers, or degrees Fahrenheit. The only competitors are the "oersted" and the "weber per square meter," and while the specialist may find these terms useful, that is no reason to confuse the issue at this point. The only significant question is: How much is a gauss? In principle, a magnetic field is measured by the amount of twist it exerts on a compass needle. The earth's field is somewhat less than half a gauss. An ordinary compass needle is useful for measuring fields of just this order of magnitude–a quarter of a gauss, a tenth of a gauss–or, on the ascending scale, two gauss, or ten gauss. Smaller fields do not exert measurable torques on compass needles. Other means for measuring them must be found. Stronger fields change the magnetization of the compass needle. More firmly fixed magnets must be found, such as atoms or atomic nuclei.

The range of magnetic fields with which one usually works in a laboratory extends from the earth's field up to 30 or 40,000 gauss. Magnets with iron cores can produce fields of the order of the higher limit. Electric currents magnetize the iron, which is a most valuable material–up to a point. Beyond this, it is a nuisance. There is a limit to the fields it can produce. Above the limit the action of electric currents alone, currents driven by powerful generators, can be more effective.

This limitation on the usefulness of iron goes back

to an essential property of magnets. If you let two bar magnets come together, they will tend to adopt the configuration shown in Fig. 52a. They will cancel each other. It is possible for them to line up in the same direction, as in Fig. 52b, but if they are

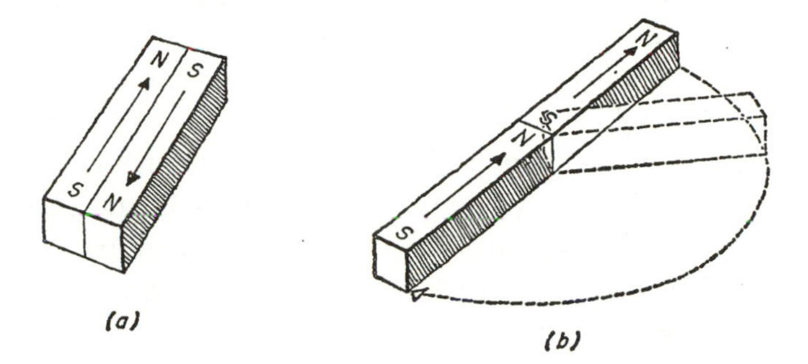

Fig. 52.

slightly displaced, as indicated by the dotted lines, the separated north and south poles will attract each other, and will swing into the position shown in Fig. 52a. A similar condition exists in atoms. Each electron, as it circles around the nucleus, produces effects like those of a magnet. Pairs of electrons tend to cancel out magnetically as in Fig. 52a. But atoms with an odd number of electrons tend to be magnetic.

Magnetic atoms also tend to cluster together in such a way that their magnetism is canceled. Hydrogen atoms are little magnets, but hydrogen molecules, consisting of two atoms each, are not magnetic because they are stuck together as in Fig. 52a. In general, this rule holds. Almost all solids and liquids are non-magnetic, partly because the individual magnetic particles tend to stick together in a way that cancels their respective magnetic properties. There are exceptions, which have been most useful in our study of other atomic forces. For example, the gas oxygen is composed of two atoms with eight electrons each, and yet this gas is one of the very few that consist of magnetic molecules. But as we investigate matter as we find it in nature or build it up in the chemistry or metallurgy laboratory, we find that it is essentially non-magnetic. Wood, water, grass, copper, leather, blood, paper, lead, oil, earth, linen, wool, glue—all these substances are hardly affected by a magnet. The big and startling exceptions are iron and neighboring elements in the periodic table—nickel and cobalt—and a few alloys and chemical compounds of elements chiefly in this part of the periodic table. The same forces that make the electrons in O_2 line up as in Fig. 52a make the electrons in the ferromagnetic elements tend to line up. Compared with ordinary non-magnetic matter, iron and steel are strongly magnetic.

But there is a definite ceiling to what can be done with iron and related ferromagnetic substances. They can be used to produce strong fields up to a point; we must surround with suitably oriented atomic magnets the point at which a strong field is wanted. But when the atomic magnets are lined up, we are through. We cannot change the intrinsic magnetization of the atoms.

This is where new techniques are needed. Where do magnetic fields come from? Can we boost the fields of magnets by going back to fundamentals? The answer is, of course, yes. Ever since Hans Christian Oersted, almost 150 years ago, discovered that a compass needle could be deflected by an electric current, scientists have realized that when large currents pass through a coil of wire, large fields are produced at its center. The larger the current, the larger the field, with no limit in sight—at least insofar as this part of the proposition goes. The added point to be considered is: How much current can you actually put through a loop or coil of wire? This is the practical question which we shall take up. Iron magnets, as we said, are excellent—up to a point. In practice, this maximum point is in the range of 30 to 40,000 gauss, depending on the size, or extent in cubic centimeters, of the space in which the field is needed. Beyond, we must consider other means; as a matter of fact, all that is left is to send larger and larger currents through a coil of wire.

The problem which I hope to make interesting and challenging to the reader is to settle some points about this simple proposition. How can you decide just how far it is possible to go? And how could one go further than is at present feasible? In order to settle this, we shall have to examine, with very considerable care and attention to detail, the simple facts about currents flowing through electrical conductors.

The Fields of Coils

Let us take up the design of a small air core magnet, which is simply a coil of wire on a non-magnetic spool, to be operated from a storage battery. How great a field can we hope to produce? How large a coil should we build? What size wire should we use? How fast will it heat up? How long will the field last before the battery runs down? These are some of the questions that must be answered before we can feel that we know what we are doing.

As a starting point we must go back to Oersted's

discovery and formulate in quantitative terms the magnetic field in the vicinity of a current. This is simply the result of careful observation. It is given in terms of a current element, as in Fig. 53. If we

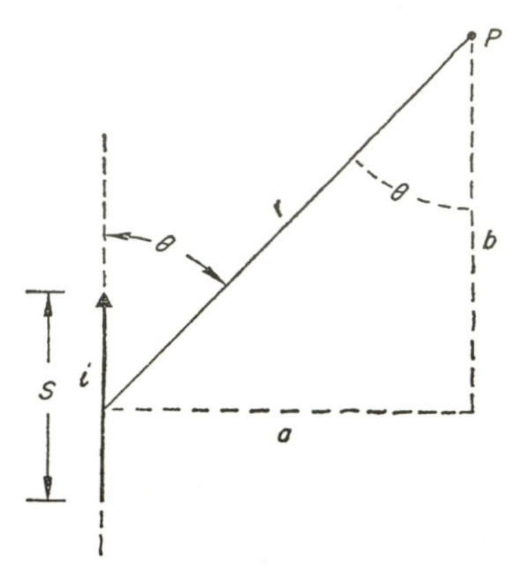

Fig. 53. The field at P *due to the length* s *of a conductor carrying a current is downward into the paper for the conditions shown. The field is in the form of circular rings around the current as axis.*

consider a small length of a current carrying wire, the field which it produces at any point P is:

$$H = k\frac{is}{r^2}\sin\,\theta$$

It is proportional to the strength i of the current, and to the lengths of the segment we are considering. It is inversely proportional to the square of the distance from the current element, and it varies with the angle between the direction of the current and the direction of the line from the current element to P. k is a number which settles units. If we wish the field to be expressed in gauss, and if i is in amperes, and r in centimeters, then k is 1/10. The field at the center of a circular loop of current is especially easy to derive from this, as may be seen by referring to Fig. 54. The field due to every current element, or

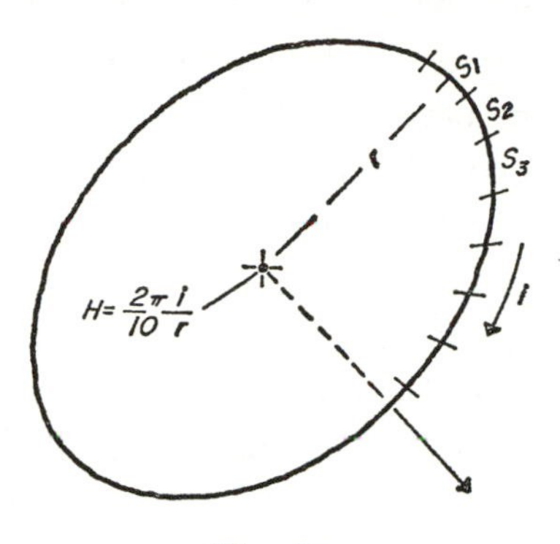

Fig. 54.

piece of current carrying conductor of length s, is the same as that of every other such element, and the resultant field of all the elements is found by adding them all together. In this case r and i are at right angles to each other, so $\sin\theta = 1$ since $\sin 90° = 1$. The resultant field is then the sum of the fields produced by the various current elements, $s_1, s_2, s_3, \ldots$

$$H = \frac{1}{10}\frac{is_1}{r^2} + \frac{1}{10}\frac{is_2}{r^2} + \ldots$$

Factoring out the common factor $\frac{1}{10}\frac{i}{r^2}$, this reduces to

$$H = \frac{1}{10}\frac{i}{r^2}(s_1 + s_2 + s_3 + \ldots)$$

But the sum $s_1 + s_2 + s_3 \ldots$ is the circumference of the circle, or $2\pi r$, so that

$$H = \frac{1}{10}\frac{i \cdot 2\pi r}{r^2} = \frac{2\pi}{10}\frac{i}{r} \text{ gauss}$$

To give an idea of magnitudes, we see that for a circular loop 1 cm in radius carrying a current of one ampere, the field at the center is .6 gauss, or a bit more than the earth's field.

This indicates the problems ahead. The *Handbook of Chemistry and Physics* gives us the allowable current-carrying capacity of a bare copper wire 1 mm in diameter as 10 amps. With this current we could make a field of about 6 gauss, in a loop nearly an inch in diameter. This is not much compared to the thousands of gauss that can be generated with an iron core magnet. But how about more turns? In Fig. 55 is a sketch of how a coil would look if it con-

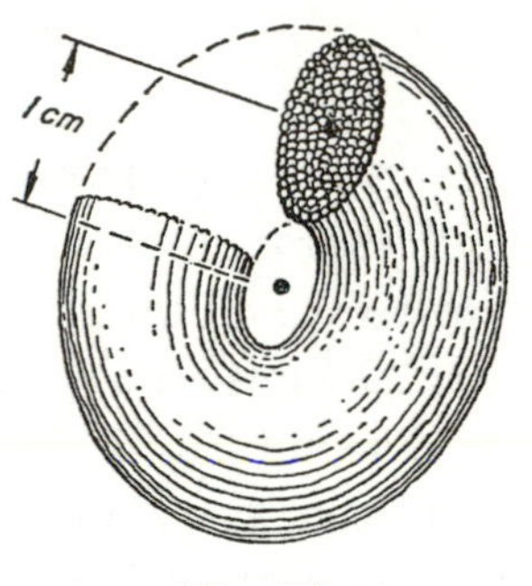

Fig. 55.

tained about 100 turns similar to the single turn we have been studying. With this coil we could make about 60–70 gauss at the center per ampere of current. How much voltage would it take to drive 1 amp through the coil? Again, using the *Handbook,* we find that the resistance of copper wire 1 mm in diameter at room temperature is about .02 ohm per meter of length. The length of a single turn is, on the average, $2\pi r$, or 6.28 cm, or for 100 turns, 6.28 meters, so that the total resistance is $6.28 \times .02$ or about .13 ohm. Using Ohm's law, which states that the voltage required to drive a current i through a resistance R is simply the product iR, we find the required voltage to be

$$V = iR = 1 \times .13 = .13 \text{ volt}$$

This is a very low voltage. A 12-volt storage battery could push one hundred times as much current through this coil, and would produce 100 times as much field, or $100 \times 60 = 6000$ gauss! How fast would the coil heat up? How long would the battery last? Could we push this number up a little more? A lot more?

Here we must emphasize an important point. We have been going at this problem in an amateurish way. We are not in a position to answer these questions except after quite lengthy calculations. And if in the end we find something not quite satisfactory, we must start all over again. And when we have done that, we may begin to wonder whether a different shape, or a different arrangement in some other respect, might not be better.

By means of a quite different approach we can do all these things at once. We can make graphs which give us facts covering many different coils. We can derive formulas which reduce the number of calculations. This is the procedure a coil designer would want to follow.

Variables in Designing a Coil

The most important decision to be made concerns the "language," or more precisely, the variables with which we want to describe our problems. Perhaps the biggest psychological difficulty is to get rid of all ideas of amperes and wires. Our initial formula for a magnetic field, to be sure, depends on the idea of electrical current. But forget that for a moment. We want to design a coil which will have a certain size and weight. It will be made of some conducting material, presumably copper, and some sort of insulation. When we connect it to a battery, it will produce a certain field and it will heat up at a certain rate. If the rate of heating is to be a limiting factor, and if the electrical power input is responsible for the heating, we will want to relate both the temperature rise and the magnetic field which we can generate to the power input.

The first of these is purely a matter of definition. The specific heat of a material, for instance copper, is a quantity we can look up in the *Handbook*. The table tells how many calories are required to raise the temperature of one gram of the substance by one degree centigrade. For a starter, while we are at it, let us list the properties of a variety of metals out of which we might build our magnet—copper, silver, gold, aluminum, lead, iron, nickel, and tin. A variety of properties of these substances are given in the following table.

The field produced by a coil of wire depends on the size and shape of the coil, and of course on the current flowing through the conductor. It is the size of the wire that is important in considering the heating or cooling of the coil, and we must know how much heating energy is supplied to it in every unit of volume. In any electrical circuit a fundamental relationship is that the heating energy supplied to a resistor per second when it is carrying a current i is i^2R watts, or joules per second. In order to find out how fast the temperature of an electric circuit element rises we must translate the input power i^2R watts into calories per second. Then, knowing the

PROPERTIES OF VARIOUS METALS, RELATIVE TO MAGNET DESIGN

	k		D	ρ	$h = \frac{1}{4.2\ kD}$
	specific heat in cal/gm°C		*density*	*resistivity ohm-cm*	*temp. rise in °C/sec for an input of 1*
	20°C	100°C	*gm/cc*	*at* 100°C	*watt/cc*
Aluminum (Al)	.021	.022	2.7	3.86×10^{-6}	4.0
Copper (Cu)	.092	.094	8.92	2.28×10^{-6}	.285
Gold (Au)	.031	.031	19.3	2.97×10^{-6}	.40
Iron (Fe)	.107	.115	7.86	16.6×10^{-6}	.265
Lead (Pb)	.031	.032	11.34	27.8×10^{-6}	.66
Nickel (Ni)	.105	.115	8.90	10.3×10^{-6}	.233
Silver (Ag)	.056	.056	10.5	2.15×10^{-6}	.41
Tin (Sn)	.054	.057	7.28	19.0×10^{-6}	.57

volume of copper in the coil and its density, we must compute the weight of the coil. And finally, knowing the specific heat, or the number of calories required to raise the temperature of 1 gram of copper by 1°, we can calculate the rate of temperature rise in the coil for any given power input. So here is a little unavoidable drudgery, the sort of thing one must learn to do reliably the first time, and yet quickly. We want to know a number *h,* the rate of heating of our various metals when we dissipate in them 1 watt in each cubic centimeter. By definition, the temperature change when we supply *k* calories to a gram of substance is 1°C. If the density, or the number of grams per cc, is D, then we must supply kD calories per second to each cc in order to raise its temperature at the rate of 1°C. per second. But 1 cal/sec = 4.2 watts (see the *Handbook*). The required quantity, then, is: h = rate of heating when we dissipate 1 watt/cc =

$$\frac{1}{4.2\ kD}$$

Let us take the specific heats at 100°C. as representative. We list in the last column of the table the quantity h, which we shall need. Finally, we shall want information about the current-carrying capacity of these various metals. The electrical resistance of a wire is proportional to its length (two equal wires in series have twice the resistance of either one), and inversely proportional to its area (two equal wires side by side, or in parallel, have half the resistance of a single one). In comparing the current-carrying capacities of different materials, we make use of the "resistivity" which is the resistance of a cube, 1 cm on a side. This resistivity ρ (Greek letter rho) can be used to determine the resistance R of any wire of length L and cross-sectional area A simply as follows:

$$R = \frac{\rho L}{A}$$

Now since the energy dissipated in any resistance R is i^2R watts, and if the current in 1 cm^2 of area is j, then the power dissipated in 1 cc is

$$p \text{ (watts/cc)} = j^2\rho$$

where j is the current density or current per unit cross section of conductor, and ρ is the resistivity.

The next problem is to discover how much magnetic field we can produce with a given amount of power in a coil of given size and shape. Let us first carry this process through in an approximate fashion for a coil similar to that sketched in Fig. 55. We have seen that a current i in a single loop of radius a will produce a field

$$\frac{2\pi}{10}\frac{i}{a}$$

and we shall use the approximation that N such loops will produce N times this field, even though some turns have a smaller radius, some a larger radius, some are slightly off to one side of the center, some slightly off to the other. The field of this arrangement will be

$$\frac{2\pi Ni}{10a}$$

and in using this formula we must be careful not to let N become so large that there is no room in a coil of average radius a for the assumed number of turns. We may write Ni in terms of the current density j, or the current per unit area of copper, times the cross-sectional area A of the coil

$$H = \frac{2\pi jA}{10a}$$

We can now proceed to write the total power W dissipated in the coil in terms of the current density, the resistivity, and the volume of the coil. We have found that the number of watts/cc is $j^2\rho$, and hence

$$W = j^2\rho V$$

where V is the volume of the coil. We can then solve this expression for j, substitute this result into the expression for the field, and so obtain a relationship between field and power. But first it is useful to consider the variables in terms of which we want to describe the coils whose properties we wish to establish. As we have set the problem up, we shall get H in terms of W, ρ, a, V, and A. It would perhaps be better, at this point, to consider how we might wind a coil. A useful and easily constructed shape is a coil having a rectangular cross section, as in Fig. 56. The conventional designations of the

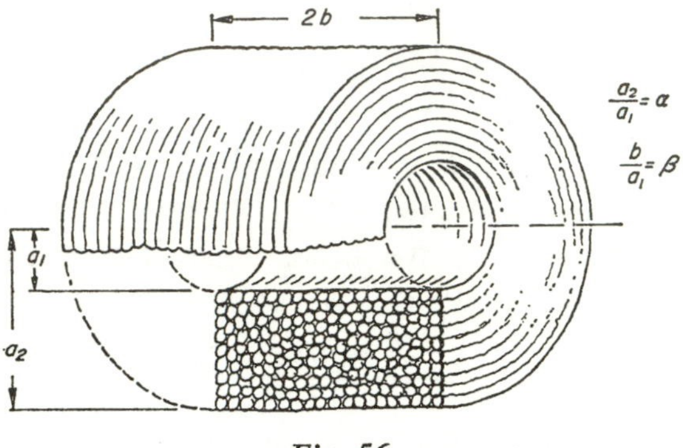

Fig. 56.

inside radius, outside radius, and length are a_1, a_2, and $2b$. In such a coil the cross-sectional area is

$$A = (a_2 - a_1)\ 2b$$

and to find the volume we shall have to multiply this area by a third dimension, which, in this case of a coil, will be its mean circumference. The mean radius is

$$a = \frac{a_2 + a_1}{2}$$

and the mean circumference is

$$2\pi a = \pi(a_2 + a_1)$$

giving the volume

$$V = (a_2^2 - a_1^2)\ 2\pi b$$

The expressions obtained by making the indicated substitutions would still be clumsy, and not as simple to use as we should like. It has been found that certain groups of coils are magnetically similar, and if we have derived the field for one, we have derived it essentially for the whole group. This fact will appear in our result if we specify a coil, not by a_2, a_1, and b, but by a new set of numbers α (alpha) = $\frac{a_2}{a_1}$, β (beta) $= \frac{b}{a_1}$ and a_1. Notice that α and β are dimensionless ratios. They fix the shape of the coil. The remaining variable, the inside radius a_1, is a scale factor, and fixes the size of the coil. The scale factor is incidentally of great practical importance in that it tells us how much room there is in the coil for experimentation. Our final expressions for A, V, and a are:

$$A = a_1^2\ (\alpha - 1)\ 2\beta$$

$$V = a_1^3\ \pi(\alpha^2 - 1)\ 2\beta$$

$$a = a_1\ \frac{\alpha + 1}{2}$$

It is a long but not difficult process to substitute these values into the expressions for the field and power, to eliminate the current density, and to get the required expression. One necessary refinement, however, is to remember that not all the coil is occupied by copper. The insulation necessarily takes up some space. If the fraction of the total cross-sectional area of the coil taken up by copper is f, and j is the current density in the copper, then

$$Ni = j\,A \cdot f;\ H = \frac{2\pi}{10}\, j\, \frac{A}{a} f$$

and the power dissipated is

$$W = j^2\ \rho\ V \cdot f$$

After making the indicated substitutions, we find that

$$H = G\sqrt{\frac{W\ f}{\rho a_1}}$$

where G is a "geometry factor," determined by the shape of the coil alone, and the square root contains only the other vital specifications—the power, the space factor, the resistivity of the conductor, and the inside radius. In our approximate calculation, the geometry factor turns out to be

$$G = \frac{\sqrt{2\pi}}{5} \sqrt{\frac{\beta}{\alpha^2 - 1}} \cdot 2\,\frac{\alpha - 1}{\alpha + 1} \text{ (approx.)}$$

The exact expression, obtained by actually summing up (integrating) the field due to the current in every part of the coil, is almost the same. Only the geometry factor is slightly different. It is

$$G = \frac{\sqrt{2\pi}}{5} \sqrt{\frac{\beta}{\alpha^2 - 1}} \log_e \frac{\alpha + \sqrt{\beta^2 + \alpha^2}}{1 + \sqrt{\beta^2 + 1}} \text{ (exact)}$$

We have now almost reached a first important vantage point for designing air core solenoids. Let us see what this amounts to. If we choose copper as the material of which our coil is to be made, we know that the rate of heating h for a liberation of 1 watt/cc is .285°C./sec, and this rate for a coil or volume V with a total power dissipation W is

$$\frac{W}{V}\,h, \text{ or } \frac{W}{V} \times .285\text{°C./sec}$$

For a coil of given shape we can compute, once and for all, G, the geometry factor. The result (using the correct formula) is plotted in Fig. 57. The maximum possible value of G in a coil with uniform current density is .179. It is obtained for $\alpha = 3$, $\beta = 2$. For a coil with an inside diameter of, say, 1 inch, G would require an outside diameter of 3 inches and a length of 2 inches. From the diagram we can read off the G-factor for coils of almost any shape. The field of any coil is then obtainable in a few short steps. Let us compute the largest field we can expect in a coil having an inside diameter $2a_1 = 1$ cm, with $f = 1$, or very thin insulation compared to

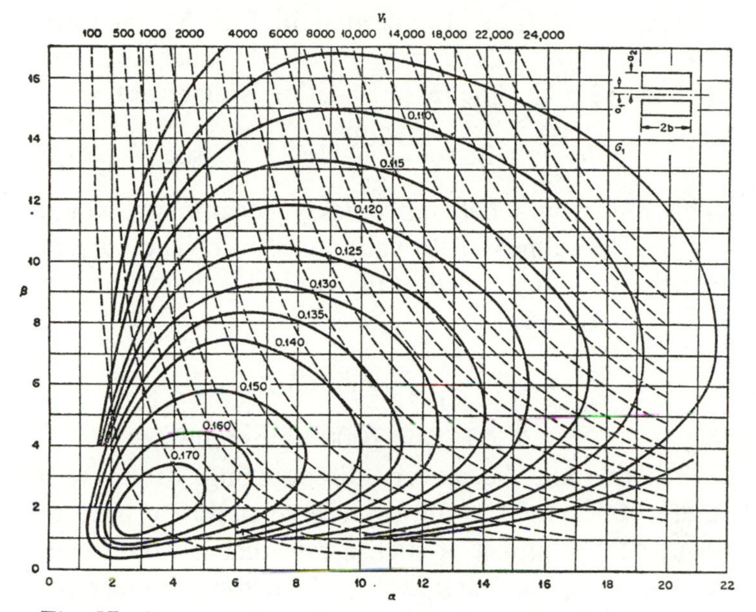

Fig. 57. Contour plots of G *and* V_1 *for coils of any shape. The first to plot* G *and* V *was the English physicist Sir John Cockcroft, celebrated for his early accelerator experiments with E. T. S. Watson.*

copper thickness, using copper with $\rho = 2 \times 10^{-6}$, and 2 six-volt batteries delivering 100 amps, or 1200 watts.

$$H = G\sqrt{\frac{Wf}{\rho a_1}}$$

$$= .179\sqrt{\frac{1200 \times 1}{2 \times 10^{-6} \times 0.5}}$$

$$= 6200 \text{ gauss}$$

a result not far different from our first design.

As for the heating, we need to know the volume, which we may write

$$V = a_1{}^3\ \pi(\alpha^2 - 1)\ 2\beta$$

$$= a_1{}^3\ V_1$$

Curves of V_1 are also plotted in Fig. 57, and we see that for $\alpha = 3$, $\beta = 2$, $V_1 = 100$, and $V = 100 (\frac{1}{2})^3 = 12.5$ cc. The rate of heating is therefore

$$h\frac{W}{V} = .285 \times \frac{1200}{12.5} \cong 28°\text{C./sec}$$

This coil would heat up quickly because we would be pouring almost 100 watts/cc into it. The current densities required to do this would be such that

$$j^2\rho = 100$$

$$j^2 = \frac{100}{2 \times 10^{-6}} = .5 \times 10^8$$

$$j = .7 \times 10^4 \text{ or } 7000 \text{ amps/cm}^2$$

All this may seem very discouraging. To make a mere 6000 gauss in an air core solenoid we have to pour energy into it at such a rate that it would melt in less than a minute! But, in fact, this is where the fun begins. Our analysis leads us to conclude that one cannot make much of a solenoid using a couple of car batteries to operate it. This is what we can *not* do. But, on the other hand, what CAN we do? We are in the position of a person who intends to write an article, and has cleared his desk, sharpened a dozen or so pencils, collected a few reference books, turned on his desk lamp, etc. We are ready to begin. Let us take a close look at our point of departure,

$$H = G\sqrt{\frac{Wf}{\rho a_1}}$$

and consider all the factors that enter into this simple expression.

First of all, there is the geometry factor. In deriving it, the assumption of a uniform current density is made. This is equivalent to saying that it is wound throughout with wire of the same size. A thorough mathematical analysis shows that small improvements are possible—in fact, very useful coils are made of flat plates with $G = .2$. But it can also be shown that no matter what you do, G can't be increased more than 50 per cent or so above the uniform current density maximum of .179. While this limitation must be borne in mind, we are not going to get from our 6000 gauss magnet up into the hundreds of thousands by changing G-factors.

Next, how about W, the power? We have been talking about putting a few kilowatts into a copper coil having a volume of around a cubic inch which would raise its temperature at the rate of about 30°C./sec. The melting point of copper is 1085°C. Actually it would take half a minute or so to melt the coil at this rate. How about more power for shorter times? Let us stick with our small coil for the moment, and see what might be done. Clearly, the more power, the shorter the time in which we can use it without melting our coil. Let us put a reasonable temperature limit into our calculations—for example, a temperature rise of 400°C. We can easily get insulation to stand this. For how long a time, which we shall call t, can we apply W watts to our coil and as a result expect the temperature to rise by an amount $T = 400$°C.? We have found h, the rate of rise of temperature of a metal when 1 watt/cc is dissipated in it. When W watts are dissipated in a coil of volume V, the rate of temperature rise will be $(W/V)h$, and the temperature rise T in t seconds will be

$$T = \frac{W}{V} ht$$

Putting in the appropriate numbers for our coil, we find that

$$400°\text{C.} = \frac{W}{12.5} .285\ t$$

$$\frac{17{,}600}{W} = t$$

and the field that could be achieved would be

$$H = .179 \sqrt{\frac{W \times 1}{2 \times 10^{-6} \times 0.5}}$$

$$= 179 \sqrt{W} \text{ gauss}$$

The conclusions to which these relations lead are tabulated below.

FIELDS THAT MIGHT BE ACHIEVED IN A COIL HAVING AN INSIDE DIAMETER OF 1 CM WITH A TEMPERATURE RISE OF 400°C.

H (*gauss*)	t (*seconds*)	W (*watts*)	Wt (*joules*)
1000	570	31	17,600
5000	23	780	17,600
10,000	5.7	3100	17,600
50,000	.23	78,000	17,600
100,000	.057	310,000	17,600
500,000	.0023	7,800,000	17,600
1,000,000	.00057	31,000,000	17,600

An inspection of this table begins to give our thinking a little scope. We can make fields ranging up to a million gauss lasting for half a millisecond in a small coil if we heat it to more than 400°C. And to do this we need 17.6 kilo-joules. Can it actually be done? Well, that much stored electrical energy is certainly not unheard of. For example, it might be done in a large bank of charged capacitors. But it takes more than that. Can we discharge such a capacitor bank through our coil in so short a time? A little more engineering analysis is needed to settle that point, but the answer is yes. Can we, then, really produce such a field? There is at least one more point to be considered, namely strength. What are the forces brought into play by the interactions of currents and magnetic fields in the coil?

When a conductor carries a current at right angles to a field, as in Fig. 58a, it experiences a force at right angles to both field and current, as shown.

$$F = 2.25 \times 10^{-7}\ Bil\ \text{lbs}$$

if B is in gauss, i in amps, and l in cm. The accurate

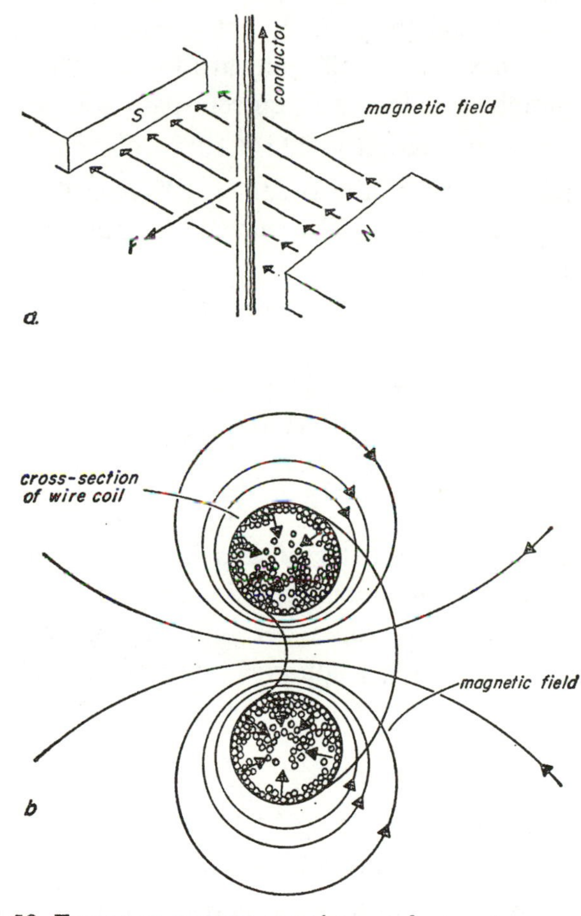

Fig. 58. Forces on current carrying conductors in a magnetic field.

calculation of the stresses in the coil shown in Fig. 58b is difficult, but if we are interested in orders of magnitude only, we can satisfy our needs fairly simply. Let us assume that all the current in the coil is flowing through a field equal to the maximum field in the center. We will be calculating too large a force, because the field actually falls off and reverses direction at some point. The forces, at right angles to the field and the current, tend to squeeze the coil together axially a little, but primarily to burst the coil. They are radially outward forces. The force per centimeter of length of current path will be

$$F \text{ lbs/cm} = .225 \times 10^{-6}\ Bi$$

The field will be of the order of 10^6 (the last case treated in the table). The current we can estimate from the total power.

$$W = j^2 \rho V = 31 \times 10^6 \text{ watts}$$

$$j^2 = \frac{31 \times 10^6}{2 \times 10^{-6} \times 12.5} = 1.25 \times 10^{12}$$

$$j = 1.1 \times 10^6 \text{ amps/cm}^2$$

$$I = j \times (a_2 - a_1)\ (2b) = 1.1 \times 10^6 \times 1 \times 2 = 2.2 \times 10^6 \text{ amps}$$

The force per centimeter of axial length of coil and per cm length of circumferential current is simply the radial bursting force per unit internal area of the coil. This force per unit area, or pressure p, is

$$p = \frac{.225 \times 10^{-6} \times 10^6 \times 2.2 \times 10^6}{2}$$

$$= .25 \times 10^6 \text{ lbs/cm}^2$$

$$= .25 \times 10^6 \times (2.54)^2 \text{ lbs/sq inch}$$

$$= 1.5 \text{ million lbs/sq inch}$$

Can we build a coil to withstand an explosive pressure of this general magnitude? (Remember that this is an overestimate of the forces, and that the actual forces will be somewhat less.) The answer, as we might expect, is that we are at the borderline. Fields of the order of half a million gauss can be contained in small coils. So far it has not been possible to hold together coils generating millions of gauss—but we no doubt will.

These few pages have shown you a little of the experimenter's art. How he must form a picture of the landscape he is exploring, for example, with the little formula

$$H = G\sqrt{\frac{Wf}{\rho a_1}}$$

and the map of G-factors in Fig. 57, and the properties of material summarized in the *Handbook*. Then he must have a mission—some particular apparatus that he wants to design for some particular set of experiments. Then he must do something that we have not touched on at all. He must design. He must draw lines on pieces of paper that forecast how machinists and metallurgists and technicians (and usually how he himself) might assemble parts for his purpose. Then he must be critical, and consider whether there are not difficulties that he had not anticipated. His work is like that of anyone who *makes* things . . . trial and error . . . try again . . . gradually gain experience . . . gradually achieve success.

Make-believe and Reality

In the foregoing pages one all-important ingredient is lacking—*reality*. The review and discussion of past explorations is an educational process, but the real meaning of education does not become apparent until it is put to use, until out of the experience of the past new ideas are formed in a fresh new mind capable of looking at existence from a slightly different point of view, eager to find the way around obstacles. In particular, the discussion of magnet design in this last chapter is hardly more than mental calisthenics. But remember that calisthenics develops habits and skills that help us to reach meaningful goals. These exercises help us to discover where to look for something new, and then actually to find the way to that something new.

Scientific authors do not like to discuss such matters in print because they don't know how to say the truth. They don't quite know. They have to guess. It seems much more worthwhile to wait until you do know, and then to describe only what you know, only the "gold" you have found, leaving out the difficulties, the descriptions of false starts, the little things that whetted the appetite and made the enterprises you were engaged on worth undertaking. So, in this set of discussions of various aspects of physics, particularly of simple quantitative mathematical aspects of physics, it may be worthwhile to end up by presenting one of many ways in which a scientist might breathe a little life into a formula like

$$H = G\sqrt{\frac{Wf}{\rho a_1}}$$

or a set of curves like those in Fig. 57.

Men and instruments are beginning to penetrate into space beyond the surface of the earth. The phenomena that occur in this high vacuum are being radioed back to us. The specific messages are clear enough: Density so and so much. Average velocity of the winds so and so much. Degree of ionization so and so much. Average particle energy so and so much. But what does that mean? What are the properties of matter under these conditions? We know matter only on our own scale of existence. Describing the condition of water in a garden hose in terms of density, chemical composition, temperature, flow, etc., would not be as useful as specifying "water." We understand this because we have lived with it—used it—all our lives. If water flows through a pipe, the character of its motion will depend on its velocity. At low velocity it will proceed by streamline motion. At higher velocities it will break into turbulent flow. Because of the extent of our experimental investigations, we can specify exactly when a velocity is "low" or "high" in this regard. The same holds true for the flow of gases, and to a much lesser extent for ionized, conducting, radiating gases called plasmas, and even less for the planetary environment which we are just beginning to penetrate. We know that magnetic fields are important. In Fig. 59 is a sketch of the lines of force of a magnetized sphere. In Fig. 60 is a sketch of the Van Allen belts of radiation, the concentrations of ionized particles around the earth. Clearly they are related to the earth's magnetic field. And in Plate V is a photograph of the solar corona, also showing the connection between radiating gases and magnetic fields.

The obvious procedure to be used in studying the phenomena is to make a theory, based on the known laws of motion and the known laws of force between charged particles. Such a theory is likely to be somewhat shaky until tested experimentally because it is found to contain many assumptions to make the

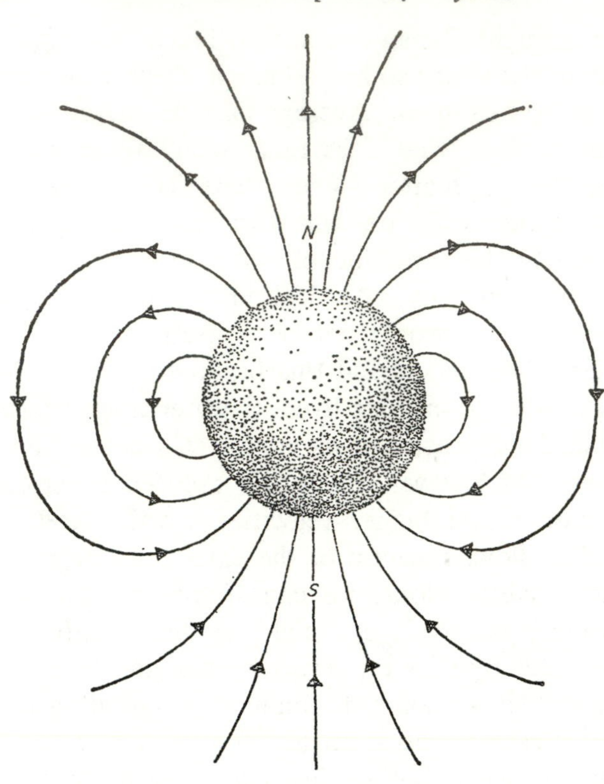

Fig. 59.

complex real situation tractable mathematically. The experimental checks are bound to be extremely expensive, and also are bound to take years and years to carry out. We cannot carry out an experiment in space as readily as we can in a laboratory.

At this point the possibility of carrying out model studies comes up. Can we put magnetic models of the earth or sun into an evacuated bottle and then reason by analogy? This is an old idea, and has been carried out more or less successfully by several researchers. The greatest difficulty is that if we attempt to reduce phenomena taking place in planetary

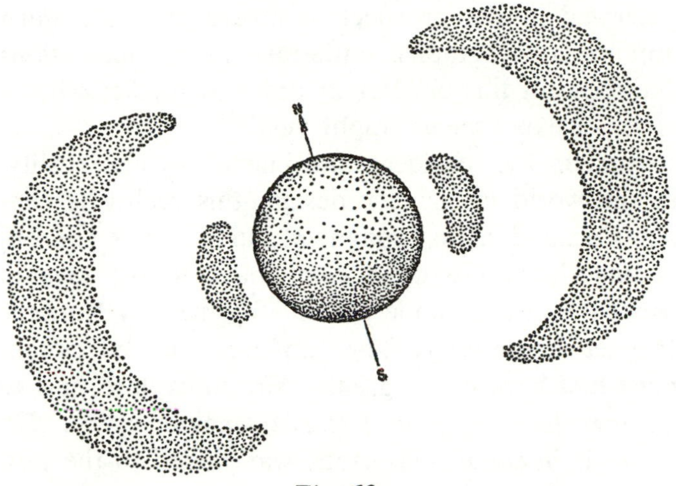

Fig. 60.

dimensions to the size of a gallon jug, we must produce a high field around the model of the earth. It must have a large magnetic moment. Some estimates of the required magnetic fields are in impossibly high intensities, millions of gauss and up.

BUT—and here is where the motivation for new work comes in—two new factors enter the picture. We are learning how to create intense magnetic fields, at least in the range of hundreds of thousands of gauss. Secondly, very large vacuum chambers have been built. To reduce auroral phenomena, or the Van Allen belts, to dimensions of the order of 10 meters rather than 10 centimeters would ease up the magnetic requirements somewhat. Do these compromises lead to experiments that are actually feasible? Undoubtedly they would be of interest, but would anticipated results justify the considerable effort and expense? And here the answer depends to a large extent on the intensity of the enthusiasm generated in the prospective investigator. He must produce many graphs, undertake many calculations like those of this chapter in order to predict what a series of experiments might yield.

And finally, the ideas and plans turn into reality. In the world of magnet design, this reality is centered around actual magnets and facilities for operating them. We must come back to the idea of new explorations made possible by the new National Magnet Laboratory. The explorers of the oceans must first have had a great desire to explore, just as we now have a great desire to explore space. The *desire* is in some mysterious way linked to the *possibility*. The desire to cross the oceans burned brighter when schemes for getting groups of ships to far distant places *and back* appeared practical. Similarly, our interest in the conditions existing in space and on the other planets is enormously heightened by the sight of the first space vehicles and the first messages which have been sent back.

But, as we have seen, the *understanding* of remote conditions, in the hydrogen atom, of the motions in our planetary system, of the simple swinging of a pendulum, is a matter related to our minds, and the kind of information that we supply to them. Magnetic model experiments may help us to gain a few insights that would take much longer to acquire from data obtained from space itself. And for the acquisition of understanding, the assembling of the essential parts of our new magnets and the machines to run them has a significance for us comparable to the sight of ships actually being constructed for Columbus and his crew, or the construction of space ships and launching facilities for today's astronauts.

We are familiar with what can be done with 10,-

000 amperes at 200 volts. There are now many laboratories with such facilities. We have been working with magnets capable of shedding this 2,000,000 watts of energy to a water-cooling system—for 25 years. But now we are building a system of copper busses capable of carrying up to 160,000 amperes at 200 volts, and a cooling system capable of carrying off this energy, and a control system capable of switching this power into any one of a dozen locations, and then controlling, or changing the operating conditions slowly or quickly, with a precision not yet attempted in such an installation.

This is the moment of birth of a new reality. Most of the world is unaware of what is happening, and rightly, for it is only a minor event in a world in which momentous events occur in rapid succession. But to a group of a few chemists, a dozen or so physicists, and some astronomers, mathematicians, geophysicists, not to mention businessmen, generals, and admirals, and others in government in a position to see distant events, this reality, carefully, quantitatively planned and finally emerging into reality from a haze of thoughts and plans, is worth watching.

Appendix

TABLE II

SOLAR AND PLANETARY LONGITUDES FROM –2500 TO +2000

The data in this table are taken from *Solar and Planetary Longitudes from —2500 to +2000,* by W. D. Stahlman and Owen Gingerich (Madison, Wisconsin: University of Wisconsin Press, 1963). The symbol meanings: *J. D.*, Julian Day; ⊙, Sun; *M,* Mercury; *V,* Venus; *J,* Jupiter; and *S,* Saturn.

			J.D.	⊙	*M*	*V*	*Mars*	*J*	*S*
1954	SEP	14	5000	171	190	217	278	114	217
1954	SEP	24	5010	181	204	226	284	116	218
1954	OCT	4	5020	191	217	233	289	117	218
1954	OCT	14	5030	200	224	238	295	118	219
1954	OCT	24	5040	210	223	241	302	119	221
1954	NOV	3	5050	220	211	240	309	119	222
1954	NOV	13	5060	230	211	235	315	119	223
1954	NOV	23	5070	241	223	229	322	119	224
1954	DEC	3	5080	251	238	225	330	119	225
1954	DEC	13	5090	261	254	226	337	118	226
1954	DEC	23	5100	271	270	231	344	117	227
1955	JAN	2	5110	281	286	237	351	116	228
1955	JAN	12	5120	292	303	246	358	115	229
1955	JAN	22	5130	302	319	255	5	113	230
1955	FEB	1	5140	312	330	265	12	112	231
1955	FEB	11	5150	322	326	276	19	111	231
1955	FEB	21	5160	332	316	287	26	110	232
1955	MAR	3	5170	342	316	298	33	110	232
1955	MAR	13	5180	352	325	310	40	110	232
1955	MAR	23	5190	2	337	321	47	110	232

			J.D.	⊙	*M*	*V*	*Mars*	*J*	*S*
1955	APR	2	5200	12	352	333	54	110	232
1955	APR	12	5210	22	10	345	61	111	231
1955	APR	22	5220	31	30	357	68	112	230
1955	MAY	2	5230	41	52	9	75	113	229
1955	MAY	12	5240	51	71	21	81	114	228
1955	MAY	22	5250	60	83	33	87	116	227
1955	JUN	1	5260	70	89	45	94	118	226
1955	JUN	11	5270	80	86	57	101	120	226
1955	JUN	21	5280	89	80	69	107	122	225
1955	JUL	1	5290	99	80	82	114	124	225
1955	JUL	11	5300	108	87	94	120	126	225
1955	JUL	21	5310	118	101	106	126	128	225
1955	JUL	31	5320	127	121	119	133	130	225
1955	AUG	10	5330	137	142	131	139	132	226
1955	AUG	20	5340	146	161	143	146	135	226
1955	AUG	30	5350	156	177	156	152	137	227
1955	SEP	9	5360	166	191	168	159	139	227
1955	SEP	19	5370	176	202	181	165	141	228
1955	SEP	29	5380	185	209	193	171	143	229
1955	OCT	9	5390	195	205	205	177	145	230
1955	OCT	19	5400	205	194	218	184	146	231
1955	OCT	29	5410	215	196	230	190	148	232
1955	NOV	8	5420	225	210	243	197	149	233
1955	NOV	18	5430	235	226	255	203	150	234
1955	NOV	28	5440	245	242	268	209	150	235
1955	DEC	8	5450	256	258	281	216	151	237
1955	DEC	18	5460	266	274	293	223	151	238
1955	DEC	28	5470	276	290	305	229	151	239
1956	JAN	7	5480	286	305	318	236	150	239
1956	JAN	17	5490	296	314	330	242	150	240
1956	JAN	27	5500	306	308	342	249	149	241
1956	FEB	6	5510	317	298	354	255	148	242
1956	FEB	16	5520	327	301	6	262	146	242
1956	FEB	26	5530	337	310	18	268	144	243
1956	MAR	7	5540	347	324	30	275	143	243

			J.D.	☉	*M*	*V*	*Mars*	*J*	*S*
1956	MAR	17	5550	357	339	41	281	142	243
1956	MAR	27	5560	7	357	52	288	142	243
1956	APR	6	5570	16	17	63	294	141	243
1956	APR	16	5580	26	38	73	301	141	243
1956	APR	26	5590	36	56	81	307	142	242
1956	MAY	6	5600	46	66	89	314	142	242
1956	MAY	16	5610	55	68	95	320	143	241
1956	MAY	26	5620	65	63	99	326	144	239
1956	JUN	5	5630	74	59	99	332	145	239
1956	JUN	15	5640	84	62	94	338	146	238
1956	JUN	25	5650	94	72	87	343	148	237
1956	JUL	5	5660	103	87	82	347	149	237
1956	JUL	15	5670	113	107	81	351	151	237
1956	JUL	25	5680	122	129	84	353	153	237
1956	AUG	4	5690	132	148	89	354	155	237
1956	AUG	14	5700	141	164	96	355	158	237
1956	AUG	24	5710	151	178	105	355	160	237
1956	SEP	3	5720	161	188	114	353	162	238
1956	SEP	13	5730	170	193	125	348	164	238
1956	SEP	23	5740	180	187	135	344	166	239
1956	OCT	3	5750	190	177	146	343	168	240
1956	OCT	13	5760	200	181	158	343	170	241
1956	OCT	23	5770	210	196	169	344	172	241
1956	NOV	2	5780	220	213	181	345	174	242
1956	NOV	12	5790	230	230	194	348	176	244
1956	NOV	22	5800	240	245	206	352	177	245
1956	DEC	2	5810	250	261	218	357	179	246
1956	DEC	12	5820	260	277	230	3	179	247
1956	DEC	22	5830	271	291	243	8	180	248
1957	JAN	1	5840	281	298	255	14	181	249
1957	JAN	11	5850	291	289	268	20	181	250
1957	JAN	21	5860	301	281	281	26	181	251
1957	JAN	31	5870	311	286	293	32	181	252
1957	FEB	10	5880	321	297	306	38	181	253
1957	FEB	20	5890	332	310	318	44	180	253

			J.D.	⊙	*M*	*V*	*Mars*	*J*	*S*
1957	MAR	2	5900	342	326	331	50	179	254
1957	MAR	12	5910	352	344	343	57	178	254
1957	MAR	22	5920	1	3	356	63	176	255
1957	APR	1	5930	11	23	8	69	174	255
1957	APR	11	5940	21	41	20	75	173	255
1957	APR	21	5950	31	49	33	82	173	255
1957	MAY	1	5960	41	48	45	88	172	254
1957	MAY	11	5970	50	41	57	94	172	253
1957	MAY	21	5980	60	39	70	101	172	253
1957	MAY	31	5990	70	45	82	107	172	252
1957	JUN	10	6000	79	56	95	113	173	251
1957	JUN	20	6010	89	73	107	119	174	250
1957	JUN	30	6020	98	93	119	125	174	249
1957	JUL	10	6030	108	115	132	132	175	248
1957	JUL	20	6040	117	134	144	138	177	248
1957	JUL	30	6050	127	151	156	145	178	248
1957	AUG	9	6060	136	163	168	151	180	248
1957	AUG	19	6070	146	173	180	157	182	248
1957	AUG	29	6080	156	175	192	163	184	248
1957	SEP	8	6090	165	168	203	170	186	249
1957	SEP	18	6100	175	160	215	176	188	249
1957	SEP	28	6110	185	167	227	183	191	250
1957	OCT	8	6120	195	183	238	189	192	250
1957	OCT	18	6130	205	201	250	196	195	251
1957	OCT	28	6140	215	217	261	202	197	252
1957	NOV	7	6150	225	233	272	209	199	253
1957	NOV	17	6160	235	249	282	215	201	254
1957	NOV	27	6170	245	264	292	222	203	255
1957	DEC	7	6180	255	277	301	229	205	257
1957	DEC	17	6190	265	282	309	236	206	258
1957	DEC	27	6200	276	272	314	243	208	259
1958	JAN	6	6210	286	265	317	250	209	260
1958	JAN	16	6220	296	272	316	257	210	261
1958	JAN	26	6230	306	284	311	264	210	262
1958	FEB	5	6240	316	298	305	271	211	263

			J.D.	⊙	*M*	*V*	*Mars*	*J*	*S*
1958	FEB	15	6250	326	314	301	278	211	263
1958	FEB	25	6260	336	331	302	286	211	264
1958	MAR	7	6270	346	350	306	293	211	265
1958	MAR	17	6280	356	9	312	300	211	265
1958	MAR	27	6290	6	25	320	307	210	266
1958	APR	6	6300	16	31	329	315	209	266
1958	APR	16	6310	26	27	339	322	207	266
1958	APR	26	6320	36	20	350	330	205	266
1958	MAY	6	6330	45	21	1	337	204	266
1958	MAY	16	6340	55	29	12	344	203	266
1958	MAY	26	6350	65	42	23	351	203	265
1958	JUN	5	6360	74	59	34	358	202	264
1958	JUN	15	6370	84	79	46	5	202	263
1958	JUN	25	6380	93	102	58	12	202	262
1958	JUL	5	6390	103	121	69	19	203	262
1958	JUL	15	6400	112	137	81	26	203	261
1958	JUL	25	6410	122	149	93	33	204	260
1958	AUG	4	6420	131	157	105	39	204	259
1958	AUG	14	6430	141	157	117	45	206	259
1958	AUG	24	6440	151	149	130	50	207	259
1958	SEP	3	6450	160	144	142	55	209	259
1958	SEP	13	6460	170	152	154	59	211	260
1958	SEP	23	6470	180	169	167	62	213	260
1958	OCT	3	6480	190	188	180	63	215	261
1958	OCT	13	6490	200	205	192	63	217	261
1958	OCT	23	6500	209	221	205	61	219	262
1958	NOV	2	6510	219	237	217	59	221	263
1958	NOV	12	6520	230	251	230	55	223	264
1958	NOV	22	6530	240	262	242	50	225	265
1958	DEC	2	6540	250	267	255	48	228	266
1958	DEC	12	6550	260	255	267	46	230	267
1958	DEC	22	6560	270	250	280	45	232	268
1959	JAN	1	6570	280	258	292	46	234	270
1959	JAN	11	6580	291	271	305	48	236	271
1959	JAN	21	6590	301	285	318	52	237	272

			J.D.	☉	*M*	*V*	*Mars*	*J*	*S*
1959	JAN	31	6600	311	301	330	56	239	273
1959	FEB	10	6610	321	318	343	60	240	274
1959	FEB	20	6620	331	336	355	64	240	274
1959	MAR	2	6630	341	355	8	69	241	275
1959	MAR	12	6640	351	9	20	75	242	276
1959	MAR	22	6650	1	13	32	80	242	276
1959	APR	1	6660	11	5	44	85	242	277
1959	APR	11	6670	21	0	56	90	241	277
1959	APR	21	6680	30	3	68	96	241	277
1959	MAY	1	6690	40	13	80	102	240	277
1959	MAY	11	6700	50	27	91	108	238	277
1959	MAY	21	6710	59	45	103	114	237	277
1959	MAY	31	6720	69	65	114	119	235	276
1959	JUN	10	6730	79	88	124	125	234	276
1959	JUN	20	6740	88	107	134	131	233	275
1959	JUN	30	6750	98	122	143	138	232	274
1959	JUL	10	6760	107	134	152	144	232	273
1959	JUL	20	6770	117	139	159	150	232	272
1959	JUL	30	6780	126	137	164	156	232	272
1959	AUG	9	6790	136	129	166	162	233	271
1959	AUG	19	6800	145	127	165	169	233	271
1959	AUG	29	6810	155	138	159	175	234	271
1959	SEP	8	6820	165	156	153	182	235	270
1959	SEP	18	6830	175	175	149	188	236	270
1959	SEP	28	6840	184	193	150	194	238	271
1959	OCT	8	6850	194	209	154	201	240	271
1959	OCT	18	6860	204	224	160	208	242	272
1959	OCT	28	6870	214	237	168	215	244	273
1959	NOV	7	6880	224	247	178	222	246	273
1959	NOV	17	6890	234	250	188	229	248	274
1959	NOV	27	6900	244	238	198	236	251	275
1959	DEC	7	6910	255	235	210	243	253	276
1959	DEC	17	6920	265	244	221	250	255	277
1959	DEC	27	6930	275	258	233	257	258	279
1960	JAN	6	6940	285	273	245	264	260	280

			J.D.	☉	*M*	*V*	*Mars*	*J*	*S*
1960	JAN	16	6950	295	289	257	272	262	281
1960	JAN	26	6960	306	306	269	279	264	282
1960	FEB	5	6970	316	323	281	287	266	283
1960	FEB	15	6980	326	341	293	294	268	285
1960	FEB	25	6990	336	354	305	301	269	286
1960	MAR	6	7000	346	355	318	309	271	287
1960	MAR	16	7010	356	346	330	317	272	287
1960	MAR	26	7020	6	342	342	325	273	287
1960	APR	5	7030	16	348	355	333	273	288
1960	APR	15	7040	25	359	7	340	274	288
1960	APR	25	7050	35	13	19	348	274	289
1960	MAY	5	7060	45	31	32	355	274	289
1960	MAY	15	7070	54	52	44	3	273	289
1960	MAY	25	7080	64	74	57	11	273	289
1960	JUN	4	7090	74	93	69	18	271	288
1960	JUN	14	7100	83	107	81	26	270	288
1960	JUN	24	7110	93	117	93	33	269	287
1960	JUL	4	7120	102	120	106	40	267	286
1960	JUL	14	7130	112	116	118	47	266	285
1960	JUL	24	7140	121	110	130	54	265	284
1960	AUG	3	7150	131	111	143	61	264	283
1960	AUG	13	7160	141	124	155	67	264	283
1960	AUG	23	7170	150	142	167	73	264	282
1960	SEP	2	7180	160	162	180	79	264	282
1960	SEP	12	7190	170	180	192	85	264	282
1960	SEP	22	7200	179	197	205	91	265	282
1960	OCT	2	7210	189	211	217	95	266	282
1960	OCT	12	7220	199	224	229	99	267	282
1960	OCT	22	7230	209	233	241	103	269	283
1960	NOV	1	7240	219	234	253	106	271	283
1960	NOV	11	7250	229	221	265	109	272	284
1960	NOV	21	7260	239	220	277	109	274	285
1960	DEC	1	7270	249	231	289	108	276	286
1960	DEC	11	7280	260	245	301	105	279	287
1960	DEC	21	7290	270	261	313	102	281	288

			J.D.	⊙	*M*	*V*	*Mars*	*J*	*S*
1960	DEC	31	7300	280	277	325	97	283	289
1961	JAN	10	7310	290	293	336	93	286	290
1961	JAN	20	7320	300	310	347	91	288	292
1961	JAN	30	7330	310	327	357	89	291	293
1961	FEB	9	7340	320	338	7	89	293	294
1961	FEB	19	7350	331	337	16	90	295	295
1961	MAR	1	7360	341	327	23	93	297	296
1961	MAR	11	7370	351	325	28	96	299	297
1961	MAR	21	7380	1	333	30	99	301	298
1961	MAR	31	7390	10	344	27	103	303	298
1961	APR	10	7400	20	0	21	108	304	299
1961	APR	20	7410	30	17	15	113	305	300
1961	APR	30	7420	40	38	11	117	306	299
1961	MAY	10	7430	49	59	13	122	307	300
1961	MAY	20	7440	59	79	17	128	307	300
1961	MAY	30	7450	69	92	24	133	307	300
1961	JUN	9	7460	78	99	32	139	307	299
1961	JUN	19	7470	88	99	41	144	307	299
1961	JUN	29	7480	97	93	51	150	306	298
1961	JUL	9	7490	107	91	61	156	304	297
1961	JUL	19	7500	116	96	72	162	303	297
1961	JUL	29	7510	126	109	83	168	301	296
1961	AUG	8	7520	135	128	95	175	301	295
1961	AUG	18	7530	145	149	106	181	299	294
1961	AUG	28	7540	155	168	118	187	298	294
1961	SEP	7	7550	164	184	130	194	298	293
1961	SEP	17	7560	174	198	142	200	297	293
1961	SEP	27	7570	184	210	154	207	297	293
1961	OCT	7	7580	194	217	166	214	297	293
1961	OCT	17	7590	204	215	178	221	298	293
1961	OCT	27	7600	214	204	191	228	299	294
1961	NOV	6	7610	224	205	204	234	300	295
1961	NOV	16	7620	234	217	216	241	301	295
1961	NOV	26	7630	244	232	229	249	303	295
1961	DEC	6	7640	254	249	242	256	305	296

			J.D.	⊙	*M*	*V*	*Mars*	*J*	*S*
1961	DEC	16	7650	264	264	254	264	307	297
1961	DEC	26	7660	275	280	267	271	309	298
1962	JAN	5	7670	285	297	279	279	311	299
1962	JAN	15	7680	295	312	292	287	313	300
1962	JAN	25	7690	305	323	304	294	315	302
1962	FEB	4	7700	315	318	317	302	318	303
1962	FEB	14	7710	325	308	329	309	320	305
1962	FEB	24	7720	335	309	342	317	323	306
1962	MAR	6	7730	345	318	354	325	326	307
1962	MAR	16	7740	355	331	7	333	328	308
1962	MAR	26	7750	5	346	19	341	330	309
1962	APR	5	7760	15	4	32	349	332	309
1962	APR	15	7770	25	24	44	357	334	310
1962	APR	25	7780	35	45	57	4	336	311
1962	MAY	5	7790	44	64	69	12	338	311
1962	MAY	15	7800	54	76	82	19	339	312
1962	MAY	25	7810	64	81	94	27	341	312
1962	JUN	4	7820	73	77	106	35	342	311
1962	JUN	14	7830	83	72	117	42	343	311
1962	JUN	24	7840	92	72	129	50	343	311
1962	JUL	4	7850	102	80	141	57	343	310
1962	JUL	14	7860	111	95	152	64	343	310
1962	JUL	24	7870	121	114	164	71	342	309
1962	AUG	3	7880	130	136	175	78	341	308
1962	AUG	13	7890	140	155	185	84	340	307
1962	AUG	23	7900	150	171	196	91	339	306
1962	SEP	2	7910	159	185	206	97	337	305
1962	SEP	12	7920	169	196	215	103	336	305
1962	SEP	22	7930	179	202	224	109	335	304
1962	OCT	2	7940	189	198	231	115	334	304
1962	OCT	12	7950	199	187	236	120	333	304
1962	OCT	22	7960	209	190	239	125	332	305
1962	NOV	1	7970	219	203	237	130	332	305
1962	NOV	11	7980	229	220	233	134	332	305
1962	NOV	21	7990	239	236	227	138	333	306

			J.D.	⊙	*M*	*V*	*Mars*	*J*	*S*
1962	DEC	1	8000	249	252	223	141	334	307
1962	DEC	11	8010	259	268	224	144	335	308
1962	DEC	21	8020	269	284	228	145	337	308
1962	DEC	31	8030	279	298	235	145	339	309
1963	JAN	10	8040	290	307	244	143	341	310
1963	JAN	20	8050	300	300	253	140	343	311
1963	JAN	30	8060	310	291	263	136	345	313
1963	FEB	9	8070	320	294	274	132	347	314
1963	FEB	19	8080	330	304	285	129	349	315
1963	MAR	1	8090	340	318	296	127	352	317
1963	MAR	11	8100	350	333	308	125	354	318
1963	MAR	21	8110	0	351	320	125	356	319
1963	MAR	31	8120	10	10	331	126	0	320
1963	APR	10	8130	20	31	343	129	2	320
1963	APR	20	8140	30	49	355	132	4	321
1963	APR	30	8150	39	59	7	135	7	322
1963	MAY	10	8160	49	60	19	139	8	322
1963	MAY	20	8170	59	54	31	144	11	323
1963	MAY	30	8180	68	51	43	149	13	323
1963	JUN	9	8190	78	54	55	153	14	323
1963	JUN	19	8200	87	65	68	159	16	323
1963	JUN	29	8210	97	80	80	164	17	323
1963	JUL	9	8220	106	100	92	170	19	322
1963	JUL	19	8230	116	123	105	175	19	322
1963	JUL	29	8240	125	142	117	181	20	321
1963	AUG	8	8250	135	158	129	188	20	320
1963	AUG	18	8260	145	171	142	194	20	319
1963	AUG	28	8270	154	181	154	200	19	318
1963	SEP	7	8280	164	186	166	207	18	317
1963	SEP	17	8290	174	180	179	213	17	317
1963	SEP	27	8300	184	171	191	220	16	316
1963	OCT	7	8310	193	175	204	227	14	316
1963	OCT	17	8320	203	190	216	234	13	316
1963	OCT	27	8330	213	207	229	241	11	316
1963	NOV	6	8340	223	224	241	248	10	316

			J.D.	⊙	*M*	*V*	*Mars*	*J*	*S*
1963	NOV	16	8350	233	240	254	256	9	317
1963	NOV	26	8360	244	256	266	263	9	317
1963	DEC	6	8370	254	271	279	271	9	318
1963	DEC	16	8380	264	284	291	279	9	319
1963	DEC	26	8390	274	292	304	286	10	320
1964	JAN	5	8400	284	283	316	294	11	321
1964	JAN	15	8410	294	274	328	302	12	322
1964	JAN	25	8420	305	280	340	309	13	322
1964	FEB	4	8430	315	291	352	317	15	323
1964	FEB	14	8440	325	305	4	325	16	325
1964	FEB	24	8450	335	320	16	333	18	326
1964	MAR	5	8460	345	338	28	341	21	328
1964	MAR	15	8470	355	357	39	349	23	329
1964	MAR	25	8480	5	17	50	357	26	330
1964	APR	4	8490	15	34	61	5	28	330
1964	APR	14	8500	24	42	71	12	30	331
1964	APR	24	8510	34	39	80	20	33	332
1964	MAY	4	8520	44	33	87	27	35	333
1964	MAY	14	8530	53	31	94	35	38	333
1964	MAY	24	8540	63	38	97	43	40	334
1964	JUN	3	8550	73	50	97	50	42	335
1964	JUN	13	8560	82	66	92	57	45	335
1964	JUN	23	8570	92	87	85	64	47	335
1964	JUL	3	8580	101	109	80	71	49	335
1964	JUL	13	8590	111	128	79	78	50	335
1964	JUL	23	8600	120	144	82	85	52	334
1964	AUG	2	8610	130	157	87	92	53	333
1964	AUG	12	8620	140	166	94	98	54	332
1964	AUG	22	8630	149	168	103	105	55	331
1964	SEP	1	8640	159	161	113	112	56	330
1964	SEP	11	8650	169	154	123	118	56	329
1964	SEP	21	8660	178	160	133	124	56	329
1964	OCT	1	8670	188	177	145	130	56	328
1964	OCT	11	8680	198	195	156	136	55	328
1964	OCT	21	8690	208	212	168	141	54	328

			J.D.	⊙	*M*	*V*	*Mars*	*J*	*S*
1964	OCT	31	8700	218	228	180	147	53	328
1964	NOV	10	8710	228	243	192	152	51	328
1964	NOV	20	8720	238	258	204	157	49	328
1964	NOV	30	8730	248	270	216	161	48	329
1964	DEC	10	8740	258	276	229	166	47	329
1964	DEC	20	8750	269	265	241	170	46	330
1964	DEC	30	8760	279	258	253	173	46	331
1965	JAN	9	8770	289	265	266	176	46	332
1965	JAN	19	8780	299	278	279	178	46	332
1965	JAN	29	8790	309	292	291	178	46	333
1965	FEB	8	8800	319	308	304	177	47	335
1965	FEB	18	8810	329	325	316	174	48	336
1965	FEB	28	8820	340	343	329	172	49	337
1965	MAR	10	8830	350	2	341	168	51	339
1965	MAR	20	8840	360	18	354	164	53	340
1965	MAR	30	8850	9	24	6	161	55	341
1965	APR	9	8860	19	18	18	159	57	342
1965	APR	19	8870	29	12	31	158	59	343
1965	APR	29	8880	39	13	43	158	61	344
1965	MAY	9	8890	48	22	55	161	64	344
1965	MAY	19	8900	58	35	68	164	66	345
1965	MAY	29	8910	68	52	80	167	69	346
1965	JUN	8	8920	77	72	93	171	71	346
1965	JUN	18	8930	87	95	105	175	73	347
1965	JUN	28	8940	96	114	117	180	75	347
1965	JUL	8	8950	106	130	130	185	78	347
1965	JUL	18	8960	115	142	142	191	80	347
1965	JUL	28	8970	125	149	154	196	82	347
1965	AUG	7	8980	134	149	166	202	84	346
1965	AUG	17	8990	144	141	178	208	86	345
1965	AUG	27	9000	154	137	190	215	87	345
1965	SEP	6	9010	163	146	202	221	89	344
1965	SEP	16	9020	173	163	213	228	89	342
1965	SEP	26	9030	183	182	225	235	90	341
1965	OCT	6	9040	193	199	236	241	91	341

			J.D.	⊙	*M*	*V*	*Mars*	*J*	*S*
1965	OCT	16	9050	203	216	248	249	91	340
1965	OCT	26	9060	213	231	259	256	91	340
1965	NOV	5	9070	223	244	270	263	91	340
1965	NOV	15	9080	233	256	280	271	90	340
1965	NOV	25	9090	243	260	290	278	89	340
1965	DEC	5	9100	253	248	299	286	88	340
1965	DEC	15	9110	263	243	307	294	86	341
1965	DEC	25	9120	273	252	312	302	84	342
1966	JAN	4	9130	284	265	315	309	83	342
1966	JAN	14	9140	294	280	314	317	82	343
1966	JAN	24	9150	304	296	309	325	81	344
1966	FEB	3	9160	314	312	302	333	81	345
1966	FEB	13	9170	324	330	298	341	81	346
1966	FEB	23	9180	334	348	299	349	81	347
1966	MAR	5	9190	344	3	303	357	82	349
1966	MAR	15	9200	354	6	310	4	82	350
1966	MAR	25	9210	4	357	318	12	83	351
1966	APR	4	9220	14	352	327	20	84	352
1966	APR	14	9230	24	357	337	27	86	353
1966	APR	24	9240	34	7	348	35	88	354
1966	MAY	4	9250	43	21	359	42	90	355
1966	MAY	14	9260	53	38	10	50	92	356
1966	MAY	24	9270	63	59	21	57	94	357
1966	JUN	3	9280	72	81	32	64	96	358
1966	JUN	13	9290	82	100	44	71	98	358
1966	JUN	23	9300	91	115	56	78	100	359
1966	JUL	3	9310	101	127	67	85	102	359
1966	JUL	13	9320	110	131	79	91	105	359
1966	JUL	23	9330	120	129	91	98	107	359
1966	AUG	2	9340	129	121	103	105	109	359
1966	AUG	12	9350	139	121	115	112	111	359
1966	AUG	22	9360	149	131	128	118	114	358
1966	SEP	1	9370	158	149	140	124	116	358
1966	SEP	11	9380	168	169	153	131	117	357
1966	SEP	21	9390	178	187	165	137	119	356

			J.D.	⊙	*M*	*V*	*Mars*	*J*	*S*
1966	OCT	1	9400	188	204	178	143	121	355
1966	OCT	11	9410	198	218	190	149	122	354
1966	OCT	21	9420	207	231	203	155	123	353
1966	OCT	31	9430	217	241	215	161	124	353
1966	NOV	10	9440	228	243	228	167	124	352
1966	NOV	20	9450	238	231	240	172	124	352
1966	NOV	30	9460	248	228	253	178	124	352
1966	DEC	10	9470	258	238	265	183	124	353
1966	DEC	20	9480	268	252	278	188	123	353
1966	DEC	30	9490	278	267	290	193	122	354
1967	JAN	9	9500	289	283	303	198	120	354
1967	JAN	19	9510	299	300	316	202	119	355
1967	JAN	29	9520	309	317	328	205	118	356
1967	FEB	8	9530	319	334	341	209	116	357
1967	FEB	18	9540	329	347	353	212	115	358
1967	FEB	28	9550	339	347	6	214	115	359
1967	MAR	10	9560	349	338	18	214	114	0
1967	MAR	20	9570	359	334	30	213	114	2
1967	MAR	30	9580	9	341	42	211	114	3
1967	APR	9	9590	19	352	54	208	115	4
1967	APR	19	9600	29	7	66	204	116	6
1967	APR	29	9610	38	24	78	200	117	6
1967	MAY	9	9620	48	45	90	197	118	7
1967	MAY	19	9630	58	67	101	196	119	8
1967	MAY	29	9640	67	86	112	195	121	9
1967	JUN	8	9650	77	100	122	195	123	11
1967	JUN	18	9660	86	110	132	198	125	11
1967	JUN	28	9670	96	111	142	201	127	12
1967	JUL	8	9680	105	107	150	205	128	12
1967	JUL	18	9690	115	102	157	210	131	13
1967	JUL	28	9700	124	104	162	215	133	13
1967	AUG	7	9710	134	117	164	220	135	12
1967	AUG	17	9720	144	136	162	226	138	12
1967	AUG	27	9730	153	156	157	232	140	12
1967	SEP	6	9740	163	174	151	238	142	11

			J.D.	⊙	*M*	*V*	*Mars*	*J*	*S*
1967	SEP	16	9750	173	191	147	245	144	10
1967	SEP	26	9760	183	205	148	251	146	9
1967	OCT	6	9770	192	218	152	258	148	8
1967	OCT	16	9780	202	226	158	265	149	7
1967	OCT	26	9790	212	226	167	272	151	6
1967	NOV	5	9800	222	214	176	280	152	6
1967	NOV	15	9810	232	213	186	287	154	5
1967	NOV	25	9820	242	224	196	295	154	5
1967	DEC	5	9830	253	240	208	303	155	5
1967	DEC	15	9840	263	255	219	310	156	5
1967	DEC	25	9850	273	271	231	318	156	5
1968	JAN	4	9860	283	287	243	326	155	6
1968	JAN	14	9870	293	304	255	334	155	7
1968	JAN	24	9880	304	320	267	342	154	7
1968	FEB	3	9890	314	332	279	350	153	8
1968	FEB	13	9900	324	329	291	357	152	9
1968	FEB	23	9910	334	319	304	5	150	10
1968	MAR	4	9920	344	318	316	12	149	11
1968	MAR	14	9930	354	326	328	20	148	12
1968	MAR	24	9940	4	338	341	27	147	13
1968	APR	3	9950	14	353	353	35	146	15
1968	APR	13	9960	23	11	5	42	145	16
1968	APR	23	9970	33	31	18	50	146	17
1968	MAY	3	9980	43	53	30	57	146	18
1968	MAY	13	9990	53	72	42	63	146	20
1968	MAY	23	0000	62	85	55	70	147	21
1968	JUN	2	0010	72	91	67	77	148	22
1968	JUN	12	0020	81	90	79	84	149	23
1968	JUN	22	0030	91	84	91	91	151	24
1968	JUL	2	0040	100	82	104	98	153	24
1968	JUL	12	0050	110	88	116	104	154	25
1968	JUL	22	0060	119	102	128	111	156	25
1968	AUG	1	0070	129	121	141	118	158	26
1968	AUG	11	0080	139	143	153	124	160	26
1968	AUG	21	0090	148	162	166	130	162	26

			J.D.	⊙	*M*	*V*	*Mars*	*J*	*S*
1968	AUG	31	0100	158	178	178	137	164	25
1968	SEP	10	0110	168	193	190	143	167	25
1968	SEP	20	0120	177	204	203	149	169	24
1968	SEP	30	0130	187	211	215	156	171	23
1968	OCT	10	0140	197	209	227	162	173	22
1968	OCT	20	0150	207	197	239	168	175	21
1968	OCT	30	0160	217	198	251	174	177	20
1968	NOV	9	0170	227	211	263	180	178	19
1968	NOV	19	0180	237	227	275	186	180	18
1968	NOV	29	0190	247	243	287	192	182	18
1968	DEC	9	0200	257	259	299	198	183	18
1968	DEC	19	0210	268	275	311	204	184	18
1968	DEC	29	0220	278	291	323	210	185	18
1969	JAN	8	0230	288	306	334	215	186	19
1969	JAN	18	0240	298	316	345	220	186	19
1969	JAN	28	0250	308	311	355	225	186	20
1969	FEB	7	0260	318	301	5	231	185	20
1969	FEB	17	0270	328	303	14	236	184	21
1969	FEB	27	0280	339	312	21	241	184	22
1969	MAR	9	0290	349	325	25	245	183	23
1969	MAR	19	0300	359	340	28	249	181	24
1969	MAR	29	0310	8	358	25	253	180	25
1969	APR	8	0320	18	17	18	256	179	27
1969	APR	18	0330	28	39	12	257	178	28
1969	APR	28	0340	38	57	9	258	177	29
1969	MAY	8	0350	47	69	11	257	177	31
1969	MAY	18	0360	57	72	15	255	176	32
1969	MAY	28	0370	67	68	22	253	176	33
1969	JUN	7	0380	76	63	30	249	176	34
1969	JUN	17	0390	86	64	39	246	177	35
1969	JUN	27	0400	95	73	49	244	178	36
1969	JUL	7	0410	105	88	60	242	179	37
1969	JUL	17	0420	114	108	70	243	180	38
1969	JUL	27	0430	124	130	81	244	182	38
1969	AUG	6	0440	133	149	93	247	183	39

			J.D.	⊙	M	V	Mars	J	S
1969	AUG	16	0450	143	165	104	250	185	39
1969	AUG	26	0460	153	179	116	255	187	39
1969	SEP	5	0470	162	189	128	261	189	39
1969	SEP	15	0480	172	195	140	267	191	39
1969	SEP	25	0490	182	191	152	273	193	38
1969	OCT	5	0500	192	181	164	279	195	37
1969	OCT	15	0510	202	183	177	286	197	36
1969	OCT	25	0520	212	197	189	293	200	35
1969	NOV	4	0530	222	214	202	300	202	34
1969	NOV	14	0540	232	231	214	307	204	33
1969	NOV	24	0550	242	246	227	314	206	32
1969	DEC	4	0560	252	262	240	322	207	31
1969	DEC	14	0570	262	278	252	329	209	31
1969	DEC	24	0580	272	292	265	337	211	31
1970	JAN	3	0590	283	300	277	344	212	31
1970	JAN	13	0600	293	293	290	352	214	32
1970	JAN	23	0610	303	284	303	359	215	32
1970	FEB	2	0620	313	288	315	6	215	32
1970	FEB	12	0630	323	298	327	13	216	33
1970	FEB	22	0640	333	312	340	21	216	34
1970	MAR	4	0650	343	327	353	28	216	35
1970	MAR	14	0660	353	345	5	35	215	36
1970	MAR	24	0670	3	4	17	42	214	37
1970	APR	3	0680	13	25	30	49	213	38
1970	APR	13	0690	23	42	42	56	212	39
1970	APR	23	0700	33	52	55	63	211	41
1970	MAY	3	0710	42	52	67	70	209	42
1970	MAY	13	0720	52	45	80	77	208	43
1970	MAY	23	0730	62	42	92	83	207	45
1970	JUN	2	0740	71	47	104	90	207	46
1970	JUN	12	0750	81	58	115	97	206	47
1970	JUN	22	0760	90	74	127	103	206	48
1970	JUL	2	0770	100	94	139	109	206	49
1970	JUL	12	0780	109	116	150	116	206	50
1970	JUL	22	0790	119	135	162	122	207	50

			J.D.	⊙	*M*	*V*	*Mars*	*J*	*S*
1970	AUG	1	0800	128	152	173	129	208	51
1970	AUG	11	0810	138	165	184	136	209	52
1970	AUG	21	0820	148	175	194	142	211	53
1970	AUG	31	0830	157	178	204	149	212	53
1970	SEP	10	0840	167	172	213	155	214	53
1970	SEP	20	0850	177	164	222	161	216	53
1970	SEP	30	0860	187	168	229	167	217	52
1970	OCT	10	0870	197	184	234	174	219	52
1970	OCT	20	0880	206	201	236	180	221	51
1970	OCT	30	0890	216	218	234	186	224	50
1970	NOV	9	0900	227	234	230	192	226	49
1970	NOV	19	0910	237	250	224	199	228	48
1970	NOV	29	0920	247	265	220	205	231	47
1970	DEC	9	0930	257	278	222	212	233	46
1970	DEC	19	0940	267	285	226	218	235	45
1970	DEC	29	0950	277	276	233	224	237	44
1971	JAN	8	0960	288	268	242	231	238	44
1971	JAN	18	0970	298	273	251	237	240	44
1971	JAN	28	0980	308	285	261	243	242	44
1971	FEB	7	0990	318	299	272	249	243	45

Index

Abstraction and symbols, 77
Acceleration: gravity and, 53, 110–11, 113; mathematical analysis of, 109–14; as slope of velocity curve, 111
Accelerators, particle, 130
Air core magnet, designing of, 137–55. *See also* Wire coils
Algebra, need for, 36
Aluminum, properties of, 143
Ammeter, use of, 29
Amplitude, definition of, 79
Angles, specification of, 92 *n*
Angstrom units (A.U.), definition of, 13
Angular momentum, definition of, 52
Archimedes, π determined by, 92–93
Asteroids, 34
Atomic nuclei: angular orientation of, 132–33; electrons and, 21; magnetic properties of, 22, 135; torques applied to, 133
Atoms: electromagnetism and, 13, 15, 17, 22, 75; magnetic fields and, 131–32; structure of, 20; vibrations of, 75
A.U. (Angstrom units), 13
Auroral phenomena, 159

Balmer, Johann, 17
Balmer series: structure of lines in, 19–20; wave lengths of lines in, 16
Bats: sonic pulse of, 84; vocal "sonar" of, 83
Beethoven, Ludwig van, 70
Billiard balls, collision of, 21
Brahe, Tycho, 28, 48
Buffon, Comte Georges Louis Leclerc de, π determined by, 94

Calculus, 36, 77; basic concepts of, 95–107; derivative in, 99; functions in, 97–107; increments in, 98; rate of change and, 102–3; slope curve in, 101
Calories, 143
Centimeter, definition of, 9
Change, rate of: acceleration and, 109–14; analysis of, 102–5. *See also* Calculus; Motion
Circular functions, 89–92
Cobalt, magnetic properties of, 136
Cockroft, Sir John, 149
Coils. *See* Wire coils

Collisions of billiard balls, 21
Colors, 76. *See also* Spectrum
Columbus, Christopher, 128–30
Computers, use of, 8
Comets, 34
Conductors, electrical, properties of, 140, 142–44, 151
Conservation of momentum, 51–52
Constants: force, 126; Planck's, 21
Constellation of stars, 37–38
Coordinate systems, 82; for star maps, 39–44
Copernicus, Nicolaus, 48
Copper: current-carrying capacity of, 140; melting point of, 151; properties of, 143; specific heat of, 142
Cosine: definition of, 90; slope of, 104–5
Cosine curves, 89–92
Cosmic waves, frequency of, 75
Creation, evidence of, 23
Current, electric. *See* Electric current
Curves: analysis of, 95–107; cosine, 89–92; frequency, 82–94; graphs for, 81, 85; rectangular, 85; sine, 89–92; slope of, 98, 100–1; triangular, 85; types of, 84; velocity, 109–11. *See also* Calculus; Graphs; Oscillations; Vibrations

Data: in astrophysics, 36–66; definition of, 25; entry of, 29–34; gathering of, 26, 28–34; kinds of, 27; raw, 28–34; for star maps, 26–66; unexpected in, 28
Density of electrical conductors, 143
Derivatives: definition of, 99; second, 103–5, 108. *See also* Calculus
Design techniques, 156; for electromagnets, 137–55
Differential calculus: basic concepts of, 95–107; derivative in, 99; functions in, 97–107; increments in, 98; rate of change and, 102–3; slope curve in, 101
Dürer, Albrecht, drawing of, 70

Earth: angular momentum of, 52; angular velocity of, 49–51; axis of rotation, 40; magnetic field of, 134, 157–58; mass of, 53; orbit of, 65; orbital radius of, 52–53; plotting revolution of, 48–54; Venus' orbit compared with, 55
Electric current: magnetizing by, 134, 145, 153; measurement of, 28–29
Electrical conductors, properties of, 140, 142–44, 151
Electromagnetic radiation: atoms and, 13, 15, 17, 22, 75; frequencies of, 75; light as, 8–9
Electromagnetism: atoms and, 13, 15, 17, 22, 25; discovery of, 136; vacuum and, 70. *See also* Magnetic fields; Wire coils
Electromagnets. *See* Wire coils
Electrons: attracted to atomic nuclei, 21; magnetic properties of, 22, 131, 135; pulsations of, 75
Ellipses, 66
Energy, 53
Equations. *See* Formulas

Experimentation, 128–61; with electromagnets, 137–55; realities in, 155–61. *See also* Data; Research

Fact gathering. *See* Data
Fields, magnetic. *See* Magnetic fields
Force: components of, 125; gravitational, 28; inverse square law of, 21; magnetic lines of, 153, 157–58
Force constant, 126
Formulas: for angular momentum of earth, 52; for Balmer series, 18; for changing velocity of particles, 82; for conversion of energy, 53; for designing electromagnets, 138–39, 141, 144–55; for determining π, 92–94; for electric power, 145; for electromagnetic properties of vacuums, 70; in Kepler's laws of planetary motion, 60–61; for laws of motion, 70, 110–11; for length of Venus year, 58–61; for line with constant unit slope, 106; for magnetic fields, 138–39, 145; Ohm's law, 141; for orbital radius of earth, 52–53; for oscillating systems, 116–23, 126–27; for periodic functions, 105–6; for planet–sun distances, 55; for rate of heating, 144; for resistivity, 144; for resolving power, 12; Rydberg, 18–22; for second derivative of curve, 103; for slope of cosine, 104; for slope of curve, 99; for slope of tangent, 99; for trigonometric functions, 90; for wave lengths radiated by hydrogen atom, 21–22
Fraunhofer lines, 13; source of, 14–15
Frequencies: definition of, 10; electromagnetic, 75; graphic representation of, 78–94; of light waves, 8, 13, 75–76; of vibrations, 79, 115
Functions: in calculus, 97–107; circular, 89–92; definition of, 80; periodic, 105–6; sine, 103–5; trigonometric, 90–92; velocity-time, 82–83

Gamma rays, 13; frequency of, 75
Gases, flow of, 157
Gauss, definition of, 134
Generalization and symbols, 77
Geometry, 36, 77, 108–14; of oscillatory motion, 114–22; of wire coils, 147–48, 150–51, 154–55. *See also* Calculus
Gingerich, Owen, 31, 34
Gold, properties of, 143
Graphs: in calculus, 98, 100–1, 104, 106, 108, 112, 116, 118; for curves, 81–85; horizontal line, 80–81; motion described by, 78–94; for rectangular curves, 85; for simple harmonic motion, 87, 89; for uniform velocities, 80–81; for wave-like motions, 82, 96
Gravity, 28, 66; acceleration due to, 53, 110–11, 113; solar, 54
Greenwich mean time, 44

Handbook of Chemistry and Physics, 140, 142, 155

Harmonic motions, 74; analysis of, 115–22; elements of, 89–94; plotting, 87–88; simple, 86–94
Heat, atomic effects of, 132
Heating, rate of, 144
Hersey, John, 69
Hexagon, circumscribed, 92–93
Hiroshima (Hersey), 69
Hydrogen atom: magnetic properties of, 135; molecules and, 135; structure of, 20; wave lengths radiated by, 21–22
Hydrogen spectrum, 15–17; wave lengths of, 16

Increments, 82; in calculus, 98–99; symbol for, 98. *See also* Calculus
Inferior conjunction, definition of, 58 *n*
Infrared radiation, 13, 18
Iron, magnetic properties of, 134, 136–37, 143
Inverse square law and electrons, 21

Julian Day, 44, 163–80
Jupiter: angular position of, 45; motion of, 47; planetary longitudes, 163–80

Kepler, Johannes, 28; difficulties faced by, 48; laws of planetary motion of, 59–61
Kinetic energy, 53
Koestler, Arthur, 45–48

Language, importance of, 67
Latitude in star maps, 41
Laws, physical: mathematical statement of, 71, 111; of motion, 70, 77, 110–11, 115, 122; inverse square law of force, 21; nature of, 19
Lead, properties of, 143
Lenses in spectroscope, 10–11
Light: bending of, 9; frequency of, 8, 13, 75–76; nature of, 8; spectrum of, 13; velocity of, 21. *See also* Electromagnetic radiation
Line, slope of, 98
Linear momentum, definition of, 52
Lines of force, magnetic, 153, 157–58
Longitude: planetary, 163–80; in star maps, 41

Magnetic fields, 134–55; atoms and, 131–33, 135–36; of earth, 134; electric current and, 145, 153; measurement of, 134, 138–39; range of, 134; solid state physics and, 133; torques applied by, 133. *See also* Electromagnetism; Wire coils
Magnetic lines of force, 153, 157–58
Magnets: design of air core, 137–56; electrons as, 22; for space exploration, 160–61. *See also* Magnetic fields
Mars: angular position of, 45; motion of, 47; orbit of, 65
Mass, 21; of celestial bodies, 53; velocity and, 20
Mathematics in physical analysis, 67–71. *See also* Calculus; Measurement
Maxwell, James, electromagnetic formulas of, 70

Measurement: in astrophysics, 36–66; of electric current, 28–29; linear, 78; of magnetic fields, 134, 138–39; significance of, 17–23; techniques of, 9; of time, 72
Mechanics. *See* Newtonian mechanics
Mercator projection in star maps, 37, 39–40
Mercury: angular position of, 45; angular separation between sun and, 61–62; length of year for, 64–65; orbit of, 63, 65; planetary longitudes, 163–80; plotting motion of, 46, 61–66
Metric system, 9
Microwaves, 13; frequency of, 75
Molecules: non-magnetic character of, 135; oxygen, 136
Momentum: definition of, 51–52; principle of conservation of, 51–52; types of, 52
Moon, 34
Motion: geometry for laws of, 108–14; graphic representation of, 77–94; laws of, 70, 77, 110–11, 115, 122; Newtonian laws of, 77, 110–11, 115, 122, 132–33; simple harmonic, 86–94, 115–22. *See also* Calculus; Oscillations; Planetary orbits; Vibrations
Musical sounds: of oboe, 84; recognition of, 73

National Magnet Laboratory, 160
Newton, Isaac, 28, 70, 77
Newtonian mechanics: gravitation in, 28; laws of motion in, 77, 110–11, 115, 122, 132–33; limitations of, 20; momentum in, 51–52
Nickel, magnetic properties of, 136, 143
Nuclei, atomic: angular orientation of, 132–33; electrons and, 21; magnetic properties of, 22, 135; torques applied to, 133

Oersted, 134
Oersted, Hans Christian, 136
Ohm's law, 140–41
Orbits, planetary, 34, 39; of earth, 48–54; elliptical nature of, 28, 66; Kepler's laws for, 59; of Venus, 54–61
Oscillations, 71–94; amplitude of, 79; characteristics of, 115–16; description of, 77–85; determining period of, 122–27; geometry of, 114–22; graphic representation of, 78–94, 116; Newton's laws of motion applied to, 77; of pendulum, 72–73, 83; repetitive cycles in, 79. *See also* Frequencies; Vibrations
Oscilloscopes, use of, 8
Order in universe, 23
Oxygen, magnetic properties of, 136

Particle accelerators, 130
Particles, 157; changes in velocity of, 82, 109–14; high-energy, 130
Perception, aids in, 7–8
Periodic functions, 105–6
Perturbations, definition of, 42
π (pi), determination of, 92–94

Planetary longitudes, 163–80
Planetary orbits, 34, 39; of earth, 48–54; elliptical nature of, 28, 66; Kepler's laws for, 59; of Venus, 54–61
Planets: angular positions of, 44–45; positions of, 31. *See also specific planets*
Pendulum: analyzing motion of, 97, 124–27; graphic representation of, 84; Newton's laws of motion applied to, 77, 132; oscillation of, 72–73, 83
Physical laws: mathematical statement of, 71, 111; inverse square law of force, 21; of motion, 70, 77, 110–11, 115, 122; nature of, 19
Physical measurement. *See* Measurement
Physics: challenge of, 2–8; role of mathematics in, 36, 67–71; solid state, 131–33
Planck's constant, 21
Polygon, circumscribed, 92–93
Potential energy, 53
Prediction: physical laws and, 71; power of, 20, 95–97
Prisms in spectroscopes, 10–11
Pythagorean theorem, 90

Radar, frequency of waves in, 75
Radians in angle, 92 *n*
Radiation. *See* Electromagnetic radiation
Radio waves, 13; frequency of, 75
Research: inspiration in, 35; techniques of, 128–31. *See also* Data; Experimentation
Resistivity: of electrical conductors, 143; formula for, 144
Resolving power, formula for, 12
Rydberg constant, definition of, 19
Rydberg formula, 18; inexactness of, 19; modifications of, 22; validity of, 20

Saturn: angular position of, 45; motion of, 47; planetary longitudes, 163–80
Scientific research: inspiration in, 35; techniques of, 128–31. *See also* Data; Experimentation
Silver, properties of, 143
Simple harmonic motion, 86–94; analysis of, 115–22; definition of, 86; elements of, 89–94; plotting, 87–88. *See also* Motion; Oscillations; Vibrations
Sine, definition of, 90
Sine curves, 89–92; analysis of, 103–5
Solar corona, 157
Sky Publishing Company, 37 *n*
Slope: constant unit, 107; of cosine, 104; of a curve, 98–107; in graphic representation of motion, 80; of a line, 98–99. *See also* Calculus
"Slope-curve," definition of, 101
Solar longitudes, 163–80
Solar system, 35; theories of, 48
Solenoids, air core, 148. *See also* Wire coils

Solids, internal constitution of, 131–33
Sounds: musical, 84; recognition of, 73
Space exploration, 160
Specific heat of electrical conductors, 142–43
Spectral lines: definition of, 10; of sunlight, 13. *See also* Spectrum
Spectroscope, 8–15; components of, 10–12; function of, 9; schematic diagram of, 11
Spectrum: definition of, 8; hydrogen, 15–17; of sunlight, 13
Star maps: data for, 26–66; Mercator projection for, 37, 39–40; preparing, 39–44; types of, 36
Stars: constellation of, 37–38; motions of, 34; positions of, 41
Steel, magnetic properties of, 136
Stoney, George, 17
Sun, 34, 35; angular separation between Mercury and, 61; angular separation between Venus and, 55–56; gravitational attraction of, 54; mass of, 53; motion of earth around, 48–53; plotting position of, 37–39, 46; solar longitudes, 163–80
Sunlight, spectrum of, 13
Superior conjunction, definition of, 58 *n*
Symbols: for increments, 98; abstraction and, 77; types of, 68

Temperature and atomic reactions, 132
Tide curve, 84
Time: displacement and, 80; measurement of, 72; velocity as function of, 82–83
Tin, properties of, 143
Tones, combining, 73; of oboe, 84
Torques, magnetic, 133
Trajectories, analysis of, 113
Trigonometry: in astrophysical calculations, 55; formulas in, 90, 92; table of functions, 91
Tables: for angles in radians, 126; for magnetic fields, 152; for natural trigonometric functions, 91; for properties of metals, 143; for sine of angles, 126; for solar and planetary longitudes, 163–80; for wave lengths of hydrogen spectrum, 16

Ultraviolet radiation, 13, 18
Universe, creation of, 23

Vacuum, electromagnetic properties of, 70
Van Allen belts, 157, 159
Velocity: angular, 49–51; mass and, 20; time and, 82–83; uniform, 80
Velocity curve, 109–11; formula for, 113
Venus: angular position of, 45; angular separation from sun, 55–56; angular separation between sun and, 55–56; earth's orbit compared with, 55; length of year for, 57–61; orbit of, 54–61, 65; plotting motion of, 46, 54–61; planetary longitudes, 163–80
Vibrations, 72; amplitude of, 79; of atoms, 75; basic

Vibrations (*cont'd*)
characteristics of, 79, 115–16; graphic representation of, 78–94; fundamental, 74; musical, 73; repetitive cycles in, 79; types of, 74, 83, 85. *See also* Frequencies; Oscillations
Vision, limitations of, 76

Wallis, John, π determined by, 93
Water, flow of, 157
Watershed, The (Koestler), 45–48
Watson, E. T. S., 149
Wave lengths: of Balmer series, 16; formulas for, 12; spectroscopic specification of, 12; of sunlight, 13; of visible light, 8–9; radiated by hydrogen atom, 21–22. *See also* Oscillations; Vibrations
Waymouth, John, 30, 32–33
Weber per square meter, 134
Wire coils, 137–55; conductors for, 142–43; contour plots, 149; designing, 141–55; forces on current-carrying, 153; geometry factor in, 147–48, 150–51, 154–55; magnetic field produced by, 144–45; power dissipation in, 145–46. *See also* Electromagnetism

X-rays, 13; experiments with, 130; frequency of, 75

A CATALOG OF SELECTED
DOVER BOOKS
IN SCIENCE AND MATHEMATICS

A CATALOG OF SELECTED

DOVER BOOKS

IN SCIENCE AND MATHEMATICS

Astronomy

BURNHAM'S CELESTIAL HANDBOOK, Robert Burnham, Jr. Thorough guide to the stars beyond our solar system. Exhaustive treatment. Alphabetical by constellation: Andromeda to Cetus in Vol. 1; Chamaeleon to Orion in Vol. 2; and Pavo to Vulpecula in Vol. 3. Hundreds of illustrations. Index in Vol. 3. 2,000pp. 6⅛ x 9¼.
23567-X, 23568-8, 23673-0 Three-vol. set

THE EXTRATERRESTRIAL LIFE DEBATE, 1750–1900, Michael J. Crowe. First detailed, scholarly study in English of the many ideas that developed from 1750 to 1900 regarding the existence of intelligent extraterrestrial life. Examines ideas of Kant, Herschel, Voltaire, Percival Lowell, many other scientists and thinkers. 16 illustrations. 704pp. 5⅜ x 8½. 40675-X

A HISTORY OF ASTRONOMY, A. Pannekoek. Well-balanced, carefully reasoned study covers such topics as Ptolemaic theory, work of Copernicus, Kepler, Newton, Eddington's work on stars, much more. Illustrated. References. 521pp. 5⅜ x 8½.
65994-1

AMATEUR ASTRONOMER'S HANDBOOK, J. B. Sidgwick. Timeless, comprehensive coverage of telescopes, mirrors, lenses, mountings, telescope drives, micrometers, spectroscopes, more. 189 illustrations. 576pp. 5⅜ x 8¼. (Available in U.S. only.)
24034-7

STARS AND RELATIVITY, Ya. B. Zel'dovich and I. D. Novikov. Vol. 1 of *Relativistic Astrophysics* by famed Russian scientists. General relativity, properties of matter under astrophysical conditions, stars, and stellar systems. Deep physical insights, clear presentation. 1971 edition. References. 544pp. 5⅜ x 8¼. 69424-0

Chemistry

CHEMICAL MAGIC, Leonard A. Ford. Second Edition, Revised by E. Winston Grundmeier. Over 100 unusual stunts demonstrating cold fire, dust explosions, much more. Text explains scientific principles and stresses safety precautions. 128pp. 5⅜ x 8½. 67628-5

THE DEVELOPMENT OF MODERN CHEMISTRY, Aaron J. Ihde. Authoritative history of chemistry from ancient Greek theory to 20th-century innovation. Covers major chemists and their discoveries. 209 illustrations. 14 tables. Bibliographies. Indices. Appendices. 851pp. 5⅜ x 8½. 64235-6

CATALYSIS IN CHEMISTRY AND ENZYMOLOGY, William P. Jencks. Exceptionally clear coverage of mechanisms for catalysis, forces in aqueous solution, carbonyl- and acyl-group reactions, practical kinetics, more. 864pp. 5⅜ x 8½.
65460-5

Math–Geometry and Topology

ELEMENTARY CONCEPTS OF TOPOLOGY, Paul Alexandroff. Elegant, intuitive approach to topology from set-theoretic topology to Betti groups; how concepts of topology are useful in math and physics. 25 figures. 57pp. $5\frac{3}{8} \times 8\frac{1}{2}$. 60747-X

COMBINATORIAL TOPOLOGY, P. S. Alexandrov. Clearly written, well-organized, three-part text begins by dealing with certain classic problems without using the formal techniques of homology theory and advances to the central concept, the Betti groups. Numerous detailed examples. 654pp. $5\frac{3}{8} \times 8\frac{1}{2}$. 40179-0

EXPERIMENTS IN TOPOLOGY, Stephen Barr. Classic, lively explanation of one of the byways of mathematics. Klein bottles, Moebius strips, projective planes, map coloring, problem of the Koenigsberg bridges, much more, described with clarity and wit. 43 figures. 210pp. $5\frac{3}{8} \times 8\frac{1}{2}$. 25933-1

CONFORMAL MAPPING ON RIEMANN SURFACES, Harvey Cohn. Lucid, insightful book presents ideal coverage of subject. 334 exercises make book perfect for self-study. 55 figures. 352pp. $5\frac{5}{8} \times 8\frac{1}{4}$. 64025-6

THE GEOMETRY OF RENÉ DESCARTES, René Descartes. The great work founded analytical geometry. Original French text, Descartes's own diagrams, together with definitive Smith-Latham translation. 244pp. $5\frac{3}{8} \times 8\frac{1}{2}$. 60068-8

THE THIRTEEN BOOKS OF EUCLID'S ELEMENTS, translated with introduction and commentary by Sir Thomas L. Heath. Definitive edition. Textual and linguistic notes, mathematical analysis. 2,500 years of critical commentary. Unabridged. 1,414pp. $5\frac{3}{8} \times 8\frac{1}{2}$. Three-vol. set.
Vol. I: 60088-2 Vol. II: 60089-0 Vol. III: 60090-4

GEOMETRY OF COMPLEX NUMBERS, Hans Schwerdtfeger. Illuminating, widely praised book on analytic geometry of circles, the Moebius transformation, and two-dimensional non-Euclidean geometries. 200pp. $5\frac{5}{8} \times 8\frac{1}{4}$. 63830-8

DIFFERENTIAL GEOMETRY, Heinrich W. Guggenheimer. Local differential geometry as an application of advanced calculus and linear algebra. Curvature, transformation groups, surfaces, more. Exercises. 62 figures. 378pp. $5\frac{3}{8} \times 8\frac{1}{2}$. 63433-7

CURVATURE AND HOMOLOGY: Enlarged Edition, Samuel I. Goldberg. Revised edition examines topology of differentiable manifolds; curvature, homology of Riemannian manifolds; compact Lie groups; complex manifolds; curvature, homology of Kaehler manifolds. New Preface. Four new appendixes. 416pp. $5\frac{3}{8} \times 8\frac{1}{2}$. 40207-X

TOPOLOGY, John G. Hocking and Gail S. Young. Superb one-year course in classical topology. Topological spaces and functions, point-set topology, much more. Examples and problems. Bibliography. Index. 384pp. $5\frac{5}{8} \times 8\frac{1}{4}$. 65676-4

LECTURES ON CLASSICAL DIFFERENTIAL GEOMETRY, Second Edition, Dirk J. Struik. Excellent brief introduction covers curves, theory of surfaces, fundamental equations, geometry on a surface, conformal mapping, other topics. Problems. 240pp. 5⅜ x 8½. 65609-8

Math–History of

A SHORT ACCOUNT OF THE HISTORY OF MATHEMATICS, W. W. Rouse Ball. One of clearest, most authoritative surveys from the Egyptians and Phoenicians through 19th-century figures such as Grassman, Galois, Riemann. Fourth edition. 522pp. 5⅜ x 8½. 20630-0

THE HISTORY OF THE CALCULUS AND ITS CONCEPTUAL DEVELOPMENT, Carl B. Boyer. Origins in antiquity, medieval contributions, work of Newton, Leibniz, rigorous formulation. Treatment is verbal. 346pp. 5⅜ x 8½. 60509-4

THE HISTORICAL ROOTS OF ELEMENTARY MATHEMATICS, Lucas N. H. Bunt, Phillip S. Jones, and Jack D. Bedient. Fundamental underpinnings of modern arithmetic, algebra, geometry and number systems derived from ancient civilizations. 320pp. 5⅜ x 8½. 25563-8

A HISTORY OF MATHEMATICAL NOTATIONS, Florian Cajori. This classic study notes the first appearance of a mathematical symbol and its origin, the competition it encountered, its spread among writers in different countries, its rise to popularity, its eventual decline or ultimate survival. Original 1929 two-volume edition presented here in one volume. xxviii+820pp. 5⅜ x 8½. 67766-4

GAMES, GODS & GAMBLING: A History of Probability and Statistical Ideas, F. N. David. Episodes from the lives of Galileo, Fermat, Pascal, and others illustrate this fascinating account of the roots of mathematics. Features thought-provoking references to classics, archaeology, biography, poetry. 1962 edition. 304pp. 5⅜ x 8½. (Available in U.S. only.) 40023-9

OF MEN AND NUMBERS: The Story of the Great Mathematicians, Jane Muir. Fascinating accounts of the lives and accomplishments of history's greatest mathematical minds–Pythagoras, Descartes, Euler, Pascal, Cantor, many more. Anecdotal, illuminating. 30 diagrams. Bibliography. 256pp. 5⅜ x 8½. 28973-7

HISTORY OF MATHEMATICS, David E. Smith. Nontechnical survey from ancient Greece and Orient to late 19th century; evolution of arithmetic, geometry, trigonometry, calculating devices, algebra, the calculus. 362 illustrations. 1,355pp. 5⅜ x 8½. Two-vol. set. Vol. I: 20429-4 Vol. II: 20430-8

A CONCISE HISTORY OF MATHEMATICS, Dirk J. Struik. The best brief history of mathematics. Stresses origins and covers every major figure from ancient Near East to 19th century. 41 illustrations. 195pp. 5⅜ x 8½. 60255-9

Physics

OPTICAL RESONANCE AND TWO-LEVEL ATOMS, L. Allen and J. H. Eberly. Clear, comprehensive introduction to basic principles behind all quantum optical resonance phenomena. 53 illustrations. Preface. Index. 256pp. 5⅜ x 8½. 65533-4

ULTRASONIC ABSORPTION: An Introduction to the Theory of Sound Absorption and Dispersion in Gases, Liquids and Solids, A. B. Bhatia. Standard reference in the field provides a clear, systematically organized introductory review of fundamental concepts for advanced graduate students, research workers. Numerous diagrams. Bibliography. 440pp. 5⅜ x 8½. 64917-2

QUANTUM THEORY, David Bohm. This advanced undergraduate-level text presents the quantum theory in terms of qualitative and imaginative concepts, followed by specific applications worked out in mathematical detail. Preface. Index. 655pp. 5⅜ x 8½. 65969-0

ATOMIC PHYSICS (8th edition), Max Born. Nobel laureate's lucid treatment of kinetic theory of gases, elementary particles, nuclear atom, wave-corpuscles, atomic structure and spectral lines, much more. Over 40 appendices, bibliography. 495pp. 5⅜ x 8½. 65984-4

AN INTRODUCTION TO HAMILTONIAN OPTICS, H. A. Buchdahl. Detailed account of the Hamiltonian treatment of aberration theory in geometrical optics. Many classes of optical systems defined in terms of the symmetries they possess. Problems with detailed solutions. 1970 edition. xv + 360pp. 5⅜ x 8½. 67597-1

THIRTY YEARS THAT SHOOK PHYSICS: The Story of Quantum Theory, George Gamow. Lucid, accessible introduction to influential theory of energy and matter. Careful explanations of Dirac's anti-particles, Bohr's model of the atom, much more. 12 plates. Numerous drawings. 240pp. 5⅜ x 8½. 24895-X

ELECTRONIC STRUCTURE AND THE PROPERTIES OF SOLIDS: The Physics of the Chemical Bond, Walter A. Harrison. Innovative text offers basic understanding of the electronic structure of covalent and ionic solids, simple metals, transition metals and their compounds. Problems. 1980 edition. 582pp. 6⅛ x 9¼. 66021-4

HYDRODYNAMIC AND HYDROMAGNETIC STABILITY, S. Chandrasekhar. Lucid examination of the Rayleigh-Benard problem; clear coverage of the theory of instabilities causing convection. 704pp. 5⅜ x 8¼. 64071-X

INVESTIGATIONS ON THE THEORY OF THE BROWNIAN MOVEMENT, Albert Einstein. Five papers (1905–8) investigating dynamics of Brownian motion and evolving elementary theory. Notes by R. Fürth. 122pp. 5⅜ x 8½. 60304-0

THE PHYSICS OF WAVES, William C. Elmore and Mark A. Heald. Unique overview of classical wave theory. Acoustics, optics, electromagnetic radiation, more. Ideal as classroom text or for self-study. Problems. 477pp. 5⅜ x 8½. 64926-1

CATALOG OF DOVER BOOKS

METHODS OF THERMODYNAMICS, Howard Reiss. Outstanding text focuses on physical technique of thermodynamics, typical problem areas of understanding, and significance and use of thermodynamic potential. 1965 edition. 238pp. 5⅜ x 8½. 69445-3

TENSOR ANALYSIS FOR PHYSICISTS, J. A. Schouten. Concise exposition of the mathematical basis of tensor analysis, integrated with well-chosen physical examples of the theory. Exercises. Index. Bibliography. 289pp. 5⅜ x 8½. 65582-2

RELATIVITY IN ILLUSTRATIONS, Jacob T. Schwartz. Clear nontechnical treatment makes relativity more accessible than ever before. Over 60 drawings illustrate concepts more clearly than text alone. Only high school geometry needed. Bibliography. 128pp. 6⅛ x 9¼. 25965-X

THE ELECTROMAGNETIC FIELD, Albert Shadowitz. Comprehensive undergraduate text covers basics of electric and magnetic fields, builds up to electromagnetic theory. Also related topics, including relativity. Over 900 problems. 768pp. 5⅝ x 8¼. 65660-8

GREAT EXPERIMENTS IN PHYSICS: Firsthand Accounts from Galileo to Einstein, edited by Morris H. Shamos. 25 crucial discoveries: Newton's laws of motion, Chadwick's study of the neutron, Hertz on electromagnetic waves, more. Original accounts clearly annotated. 370pp. 5⅜ x 8½. 25346-5

RELATIVITY, THERMODYNAMICS AND COSMOLOGY, Richard C. Tolman. Landmark study extends thermodynamics to special, general relativity; also applications of relativistic mechanics, thermodynamics to cosmological models. 501pp. 5⅜ x 8½. 65383-8

LIGHT SCATTERING BY SMALL PARTICLES, H. C. van de Hulst. Comprehensive treatment including full range of useful approximation methods for researchers in chemistry, meteorology and astronomy. 44 illustrations. 470pp. 5⅜ x 8½. 64228-3

STATISTICAL PHYSICS, Gregory H. Wannier. Classic text combines thermodynamics, statistical mechanics and kinetic theory in one unified presentation of thermal physics. Problems with solutions. Bibliography. 532pp. 5⅜ x 8½. 65401-X